Structural Equation Modeling with lavaan

Structural Equation Modeling with lavaan

Kamel Gana
Guillaume Broc

iSTE

WILEY

First published 2019 in Great Britain and the United States by ISTE Ltd and John Wiley & Sons, Inc.

ISTE Ltd
27-37 St George's Road
London SW19 4EU
UK

www.iste.co.uk

John Wiley & Sons, Inc.
111 River Street
Hoboken, NJ 07030
USA

www.wiley.com

Library of Congress Control Number: 2018956892

British Library Cataloguing-in-Publication Data
A CIP record for this book is available from the British Library
ISBN 978-1-78630-369-1

FSC www.fsc.org MIX Paper from responsible sources FSC® C013604

Contents

Preface

The core content of this book was written 20 years ago, when I began giving my first courses on structural equation modeling at the François Rabelais University in Tours, France. Having never been left pending, the manuscript has been constantly updated for the needs of my university courses and numerous introductory workshops to SEM that I was conducting in various foreign universities. These courses and workshops were both an introduction to the statistical tool and an adoption of a software without which this tool would be obscure, abstract and disembodied. To put it directly and bluntly, any introduction to structural equation modeling (SEM) compulsorily includes adopting a SEM software. Among LISREL, Amos, EQS, M*plus* Sepath/Statistica, and Calis/SAS, there is no dearth of options. These commercial programs no doubt helped in demystifying structural equation modeling and have thus given it an actual popularity that continues to grow.

Writing a book, in this case a practical handbook of structural equation modeling, requires introducing one or more of these commercial software that are admittedly quite expensive. That is where the problem lies. Not that they do not deserve it, but picking one is inevitably advertising that software. I cannot and will not consent to this. Moreover, access to these commercial software remains, for many students and young researchers, an obstacle and constraint that is often insurmountable. I have often experienced the challenge of teaching SEM in African universities where it was impossible to have SEM commercial software. I happened to use a restricted student version of a commercial program to demonstrate in my class.

When R, free open-source software, was developed, the situation changed. R is made of packages dedicated to all kinds of data analysis and processing. The lavaan package, provided by Rosseel [ROS 12], is dedicated to SEM. It achieved immediate success as it has all the features proposed by commercial software, and it offers such a disconcerting ease of use. As with any statistical tool, practice remains the best way to master SEM. There is no better way to do this than by having software at hand. R and lavaan have changed our way of teaching statistical tools as well as the way in which students can become familiarized with, adapt, and use them. Our level of demand on them changes as the students' view of these tools evolves. Understanding, learning, and especially practicing without any limits (apart from that of having a computer): this is what R and its lavaan package offer to students. This book certainly contributes to it.

Without the impetus and the decisive and meaningful contribution of Guillaume Broc, author of the book *Stats faciles avec R* (De Boeck), this manual would not have seen the light of day. We share the following belief: access to science and its tools must become popular and be available to everyone. We think that this manual, devoted to structural equation modeling with lavaan, fully contributes to this purpose.

It is because this book aims to be a didactic handbook and a practical introduction to SEM meant for students and users who do not necessarily need complex mathematical formulae to adopt this tool and be able to use it wisely, that we submitted the first draft to some novice students in SEM in order to assess its clarity and comprehensibility. Their careful reading and their judicious and pertinent comments allowed for a substantial improvement of this manual. They are very much thanked for this. In fact, they are fully involved in this project. However, we retain and accept full responsibility for mistakes, shortcomings, or inadequacies that may be present in this manual.

Kamel GANA
October 2018

Introduction

"The time of disjointed and mobile hypotheses is long past, as is the time of isolated and curious experiments. From now on, the hypothesis is synthesis."

Gaston Bachelard
Le Nouvel Esprit scientifique, 1934

"There is science only when there is measurement."

Henri Piéron, 1975

Predicting and explaining phenomena based on non-experimental observations is a major methodological and epistemological challenge for social sciences. Going from a purely descriptive approach to an explanatory approach requires a suitable, sound theoretical corpus as well as appropriate methodological and statistical tools. "It is because it is part of an already outlined perspective that structural equation modeling constitutes an important step in the methodological and epistemological evolution of psychology, and not just one of the all too frequent fads in the history of our discipline", wrote Reuchlin ([REU 95] p. 212). The point aptly described by Reuchlin applies to several other disciplines in the human and social sciences.

Since its beginning, there have been two types of major actors who have worked with structural equation modeling (SEM), sometimes in parallel to its development: those who were/are in a process of demanding, thorough, and innovative application of the method to their field of study, and those who were/are in a process of development and refinement of the method itself. The second group is usually comprised of statisticians (mathematicians, psychology psychostatisticians, etc.), whereas the first group is usually comprised of data analysts. While willingly categorizing themselves as data analysts, the authors of this book recognize the importance of the statistical prerequisites necessary for the demanding and efficient use of any data analysis tool.

This manual is a didactic book presenting the basics of a technique for beginners who wish to gradually learn structural equation modeling and make use of its flexibility, opportunities, and upgrades and extenxions. And it is by putting ourselves in the shoes of a user with a limited statistical background that we have undertaken this task. We also thought of those who are angry with statistics, and who, more than others, might be swayed by the beautiful diagrams and goodness-of-fit indices that abound in the world of SEM. We would be proud if they considered the undoubtedly partial introduction to SEM we give here as insufficient. As for those who find the use of mathematics in humain and social sciences unappealing, those who have never been convinced by the utility of quantitative methods in these sciences, it is likely that, no matter what we do, they will remain so forever. This manual will not concern them… It is hardly useful to focus on the fact that using these methods does not mean conceiving the social world or psychological phenomena as necessarily computable and mathematically formalizable systems. Such a debate has lost many a time to the epistemological and methodological evolution of these sciences. This debate is pointless…

In fact, computer software has made it possible to present a quantitative method in a reasonably light way in mathematical formulas and details. Currently, it is no longer justifiable to present statistical analysis pushed up to its concrete calculation mode, like when these calculations were by hand in the worst cases or done with the help of a simple calculator in the best cases. But the risk of almost mechanically using such programs, which often gives users the impression they are exempted from knowing the basics of technical methods and tests that they use, is quite real.

We have tried to limit this risk by avoiding making this book a simple SEM software user's guide. While recognizing the importance of prerequisite statistics essential to a demanding and efficient use of any data analysis tool, we reassure readers: less is more. Let us be clear from the outset that our point of view in this book is both methodological and practical and that we do not claim to offer a compendium of procedures for detailed calculations of SEM. We have put ourselves in the shoes of the user wishing to easily find in it both a technical introduction and a practical introduction, oriented towards the use of SEM. It is not a "recipe" book for using SEM that leads to results that are not sufficiently accurate and supported. Implementing it is difficult, as it also involves handling SEM software, thus following the logic of a user's guide.

In the first chapter following this introduction, the founding and fundamental concepts are introduced and the principle and basic conventions are presented and illustrated with simple examples. The nature of the approach is clearly explained. It is a confirmatory approach: first, the model is specified, and then tested. Handling the easy-to-learn lavaan software constitutes the content of the second chapter. Developed by Rosseel [ROS 12], the open-source lavaan package has all of the main features of commercial SEM software, despite it being relatively new (it is still in its beta version, meaning that is still in the test and construction phase). It has a remarkable ease of use.

Chapter 3 of this manual presents the main steps involved in putting a structural equation model to the test. Structural equation modeling is addressed both from the point of view of its process, that is, the different steps in its use, as well as from the point of view of its product, that is, the results it generates and their reading. Also, different structural equation models are presented and illustrated with the lavaan syntax and evaluation of the output: path models analysis and the Actor-Partner Interdependence Model (APIM). Similarly, the two constituent parts of a structural general equation model are detailed: the measurement model and the structural model. Here again, illustrations using the lavaan syntax and evaluation of the output make it possible for the reader to understand both the model and the software.

Any model is a lie as long as its convergence with the data has not been confirmed. But a model that fits the data well does not mean that it represents the truth (or that it is the only correct model, see equivalent models); it is simply a good approximation of reality, and hence a reasonable explanation of tendencies shown by our data. Allais [ALL 54] was right in writing that "*for any given level of approximation, the best scientific model is the one which is most appropriate* [italicized by the author]. In this sense, there are as many true theories as given degrees of approximation" (p. 59). Whatever it may be, and more than ever, the replication of a model and its cross-validation are required.

The fourth chapter is dedicated to what has been called "the more or less recent extensions of SEM". Here, the term "extensions" means advances and progress, because the approach remains the same, regardless of the level of complexity of the specified models and the underlying degree of theoretical elaboration. The aim here is to show the use of the power and flexibility of SEM through some examples. Its potential is immense and its opportunities multiple. Its promises are rich and exciting. However, it was not possible to go through them all. It seemed wise to focus on those that are becoming unavoidable. Moreover, some analyses have become so common that they could cease to be seen as a mere extension of basic equation models. One can think of multigroup analyses that offer the possibility to test the invariance of a model through populations, thus establishing the validity, or even universality, of the theoretical construct of which it is the representation. Latent state-trait models, which refer to a set of models designed and intended to examine stability of a construct over time (temporal), are more recent, and it is to them that we have dedicated a chapter that is both technical and practical. Finally, latent growth models that find their place in longitudinal, rare, and valuable data never cease to be of interest to researchers. Using them with the help of models combining covariance structure analysis and mean structure modeling is one of the recent advances in SEM.

We suggest that the reader acquires a progressive, technical introduction to begin with by installing the free software lavaan with no further delay. The second chapter of this book will help in getting started with this software. It is in the reader's interest to follow step-by-step the treatment of data in the book in order to replicate the models presented, and not move to the next step until they get the same results. These data will be available on a website dedicated to this manual.

We started this introduction by paraphrasing Reuchlin, We would like to conclude our introduction citing Reuchlin once again when he accurately said that SEM "are tools whose usage is not possible, it is true, unless there is some knowledge and some psychological hypotheses about the functioning of the behaviors being studied. It would be paradoxical for psychologists to consider this constraint as a disadvantage" [REU 95]. One could even say that they would be wrong to consider it in this way. And Hair, Babin, and Krey [HAI 17], marketing and advertising specialists, would agree with Reuchlin. In fact, in a recent literature review examining the use of SEM in articles published in the *Journal of Advertising* since its first issue in 1972, these authors acknowledge that the attractiveness of structural equation modeling among researchers and advertisers can be attributed to the fact that the method is proving to be an excellent tool for testing advertising theories, and they admit bluntly that the increasing use of structural equation modeling in scientific research in advertising has contributed substantially to conceptual, empirical, and methodological advances in the science of advertising.

Is it not an epistemological evolution necessary to any science worthy of the name to go from the descriptive to the explanatory? This is obviously valid for a multitude of disciplines where SEM is already used: agronomy, ecology, economy, management, psycho-epidemiology, education sciences, sociology, etc.

1

Structural Equation Modeling

Structural Equation Modeling (SEM) is a comprehensive and flexible approach that consists of studying, in a hypothetical model, the relationships between variables, whether they are measured or latent, meaning not directly observable, like any psychological construct (for example, intelligence, satisfaction, hope, trust[1]). Comprehensive, because it is a multivariate analysis method that combines the inputs from factor analysis and that of methods based or derived from multiple regression analysis methods and canonical analysis [BAG 81, KNA 78]. Flexible, because it is a technique that allows not only to identify the direct and indirect effects between variables, but also to estimate the parameters of varied and complex models including latent variable means.

Mainly of a correlational nature, structural models are both linear statistical models, whose normal distribution of variables is almost necessary, and statistical models in the sense that the error terms are considered to be partly related to the endogenous variables (meaning predicted). We say almost necessary because the success of structural equation modeling is such that its application extends, certainly with risks of error, to data obtained through categorical variables (ordinal or even dichotomous) and/or by clearly violating the multivariate normal distribution. Considerable mathematical advances (like the so-called "robust" estimation methods) have helped currently minimize these risks by providing some remedies to the non-normality of the distribution of variables and the use of data collected by the means of measurement scales other than that normally required for structural equation models, namely interval scales [YUA 00]. We will discuss more on that later.

1 See Ramirez, David and Brusco [RAM 13] for the main constructs used on marketing science.

Our first goal is to introduce the reader to the use of structural equation models and understand their underlying logic; we will not delve too much into mathematical and technical details. Here, we will restrict ourselves to introducing, by way of a reminder, the concepts of correlation, multiple regression, and factor analysis, of which structural equation modeling is both a summary and a generalization. We will provide to the reader some details about the concept of normality of distribution, meaning with linearity, a basic postulate of structural equation modeling. The reader will find the mathematical details concerning the basic concepts briefly recalled here in any basic statistical manual.

1.1. Basic concepts

1.1.1. *Covariance and bivariate correlation*

Both covariance and correlation measure the linear relationship between two variables. For example, they make it possible to learn about the relationship between two items of a test or a measure (e.g., a questionnaire) scale. Figure 1.1 provides a graphic illustration of the same.

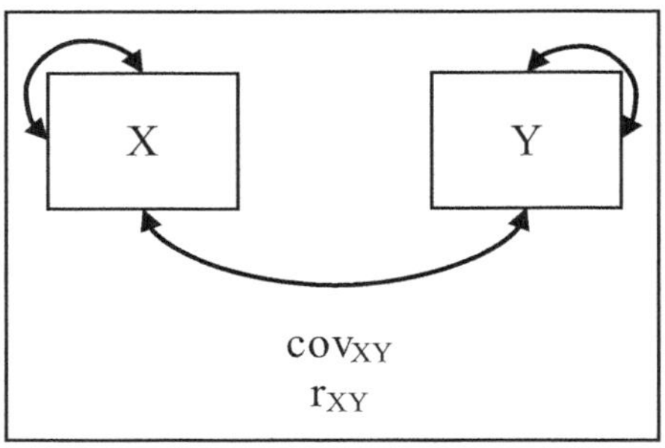

Figure 1.1. *Covariance/correlation between two variables (the small curved left-right arrow indicates the variance)*

Covariance, which measures the variance of a variable with respect to another (covariance), is obtained as follows:

$$cov_{XY} = \frac{\sum(X-M_X)(Y-M_Y)}{N-1} \quad [1.1]$$

where:

– M = mean;

– N = sample size.

Being the dispersion around the mean, the variance is obtained as follows:

$$var = \frac{\sum(X-M)^2}{N-1} \qquad [1.2]$$

The values of a covariance have no limits. Only, it should be noted that the positive values of covariance indicate that values greater than the mean of a variable are associated with values greater than the mean of the other variable and the values lesser than the mean are associated in a similar way. Negative covariance values indicate values greater than the mean of a variable are associated with values lesser than the mean of the other variable.

Unlike covariance, correlation measures such a relationship after changing the original units of measurement of variables. This change, called "standardization" or "normalization", involves centering-reducing (i.e. M = 0.00, standard deviation = 1.00) a variable (X) by transforming its raw score into z score:

$$z_X = \frac{X-M}{\sigma} \qquad [1.3]$$

where:

– M = mean of X;

– σ = standard deviation of X.

The standard deviation is simply the square root of the variance:

$$\sigma = \sqrt{var} \qquad [1.4]$$

$$= \sqrt{\frac{\sum(X-M)^2}{N-1}}$$

Remember that standard deviation is the index of dispersion around the mean expressing the lesser or higher heterogeneity of the data. Although standard deviation may not give details about the value of scores, it is expressed in the same unit as these. Thus, if the distribution concerns age in years, the standard deviation will also be expressed in the number of years.

The correlation between standardized variables X and Y (Z_X and Z_Y, [1.3]) is obtained as follows:

$$r_{XY} = \frac{\sum(Z_X)(Z_Y)}{N-1} \qquad [1.5]$$

Easier to interpret than covariance, correlation, represented, among others, by the Bravais-Pearson coefficient *r*, makes it possible to estimate the magnitude of the linear relationship between two variables. This relationship tells us what information of the values of a variable (*X*) provides information on the corresponding values of the other variable (*Y*). For example, when *X* takes larger and larger values, what does *Y* do? We can distinguish between the two levels of responses. First, the direction of the relationship: when the variable *X* increases, if the associated values of the variable *Y* tend overall to increase, the correlation is said to be positive. On the other hand, when *X* increases, if the associated values of *Y* overall tend to reduce, the correlation is called negative. Second, the strength of the association: if information of *X* accurately determines information of *Y*, the correlation is perfect. It corresponds respectively to + 1 or – 1 in our demonstration. In this case, participants have a completely identical way of responding to two variables. If information of *X* does not give any indication about values assumed to be associated with *Y*, there is complete independence between the two variables. The correlation is then said to be null. Thus, the correlation coefficient varies in absolute value between 0.00 and 1.00. The more it is closer to + 1.00 or – 1.00, the more it indicates the presence of a linear relationship, which can be represented by a straight line in the form of a diagonal[2]. On the other hand, more the coefficient goes to 0, the more it indicates the lack of a linear relationship. A correlation is considered as being significant when there is a small probability (preferably less than 5%) so that the relationship between the two variables is due to chance.

As much as the covariance matrix contains information about the relationships between the two measures (scores) and their variability within a given sample, it does not allow for comparing, unlike a correlation matrix, the strength of the relationships between the pairs of variables. The difference between these two statistics is not trivial as the discussions on the consequences of the use of one or the other on the results of the analysis of structural equation models are to be taken seriously. We will discuss this further.

Furthermore, it is worth remembering that there are other types of correlation coefficients than the one that we just saw. In structural equation modeling, the use of tetrachoric and polychoric correlation coefficients is widespread, as they are suitable for measurements other than those of interval-levels. The first is used to estimate the association between two variables, called "dichotomous"; the second is used when there are two ordinal-level variables.

---

2 The correlation is higher than the points tend to lie on the same straight line. Note that each point represents the correspondence, for example, between the scores of an individual to two items.

1.1.2. *Partial correlation*

The correspondence between two variables may result from various conditions that the calculation of correlation cannot always detect. Thus, assuming that only the well-off have the financial means to buy chocolate in a given country, even a very strong correlation observed between the two variables – "consumption of chocolate" and "life satisfaction" – in older people does not mean that the first is the cause of the second. First, we are right in thinking of the opposite, as a statistic expressed by a correlation can be read in two ways. In addition, we can think that these assumptions are all erroneous, and that it would perhaps come from a common cause, the "milieu" that similarly determines the two variables, which, due to this fact, prove to be correlated. In this case, life satisfaction in a category of the elderly does not come from consuming chocolate, but from the preferred milieu in which they live; it is this milieu that allows them to both consume chocolate and be happy. Here, we see a spurious (artificial) relationship that we will shed light on through the following illustration. Let us consider that the correlation matrix between these three variables measured from a sample of elderly people (Table 1.1).

Variable	X	Y	Z
X. Chocolate consumption	1.00		
Y. Life satisfaction	.49	1.00	
Z. Milieu	.79	.59	1.00

Table 1.1. *Correlation matrix (N = 101)*

The use of partial correlations is useful here, as it will allow us to estimate the relationship between X and Y controlling for Z that will be hold constant. This correlation is written in this way, $r_{XY.Z}$, and is calculated as follows:

$$r_{XY \cdot Z} = \frac{r_{XY} - (r_{XZ})(r_{YZ})}{\sqrt{(1 - r_{XZ}^2)(1 - r_{YZ}^2)}} \quad [1.6]$$

Considering the numerator of this equation, we can observe how the "milieu" variable (Z) was hold constant: we simply removed the two remaining relationships from the relationship between "chocolate consumption" (X) and "life satisfaction" (Y), namely (r_{XZ}) and (r_{YZ}). If we apply this formula to the data available in the matrix [1.7], the partial correlation between X and Y by controlling Z will be as follows:

$$r_{XY \cdot Z} = [0.49 - (0.79)(0.59)] / [\sqrt{(1 - 0.79^2)\,(1 - 0.59^2)} \qquad [1.7]$$

$$= [(0.49 - 0.46)] / [\sqrt{(0.37)(0.65)}]$$

$$= 0.03/0.49$$

$$= 0.06$$

We realize that by controlling for the variable "milieu", the relationship between "chocolate consumption" and "life satisfaction" fades, as it is likely an artificial one. The milieu takes the place of confounding factor, giving the relationship between "chocolate consumption" and "life satisfaction" an artificial nature, meaning spurious.

To conclude, we put emphasis on the fact that it is often unwise to interpret the correlation or partial correlation in terms of causality. Mathematics cannot tell us about the nature of the relationship between two variables. It can only tell us to what extent the latter tend to vary simultaneously. In addition, the amplitude of a link between two variables may be affected by, among other things, the nature of this relationship (i.e. linear or non-linear), the normality of their distribution, and psychometric qualities (reliability, validity) of their measures.

As for the causality, it requires three criteria (or conditions): 1) the association rule, that is the two variables must be statistically associated; 2) the causal order between variables, the (quite often) temporal order where the cause precedes the effect must be determined without ambiguity and definitely with theoretical reasons that allow for assuming the order; 3) the non-artificiality rule, in which the association between the two variables must not disappear when we remove the effects of variables that precede them in the causal order. This requires that, in the explanation of the criterion, the intermediary variable gives an additional contribution compared to the latter.

It is clear that only experimentation is able to satisfy these three criteria, within the limits of what is possible and thinkable. It goes without saying that no statistical procedure (analysis of variance, regression, path analysis), as sophisticated and clever as it may be, allows for establishing any causal order between variables. We could at most make dynamic predictions in longitudinal research [MCA 14]. In Pearl [PEA 98], we can find an original and detailed analysis of causality in SEM, and in Bollen and Pearl [BOL 13], there is a clarification of the myths about causality in SEM.

1.1.3. *Linear regression analysis*

Whether single or multiple, regression has a predictive and "explanatory" function. When studying relationships, it is true that we are right in predicting one of the variables from our knowledge of the other variable(s). We can also seek to determine the relative influence of each predictor variable. It is rather this "explanatory" aspect of regression that is best suited to structural equation modeling. Regression is in fact a linear mathematical model linking a criterion variable (to be explained) to one or more explanatory variables, a model mainly built to predict the criterion variable.

Regression is called "simple" when the model, very rarely, has only one predictor variable. Figure 1.2 represents such a model where X is the predictor variable, considered as a predictor of the criterion variable, or criterion, Y. B (β) is the regression coefficient, which can be either non-standardized (B) or standardized (β), while (e) refers to the prediction error (residual), the part of variance that is unexplained and probably due to variables ignored by the model. Regression analysis aims to estimate the value of this coefficient and establish the part of the error in the variance of the criterion. In other words, Figure 1.2 shows Y as subject to the "effects" of X and as well as the error term (e).

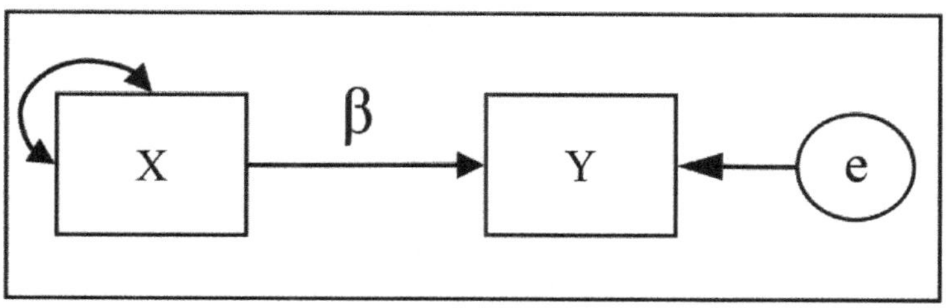

Figure 1.2. *Model of a simple linear regression (the curved left-right arrow indicates the variance)*

The model is then written as follows:

$$Y = \alpha + \beta X + e \qquad [1.8]$$

This is a regression equation describing the structural relationship between the two variables (X and Y) where α refers to Y-intercept. The estimate of the coefficient B (β) is solved using the OLS method that involves finding the values that minimize the sum of squares of the difference between the observed values and the values predicted by the linear function (we will come back to this when we discuss discrepancy functions and estimation methods).

It also allows to optimize the correlation between these two variables, considered as multiple correlation (*R*) and whose square (R^2) shows the proportion of variance of the criterion variable attributable to the explanatory variable in the case of a simple regression, or to all the explanatory (predictor) variables in the case of a multiple regression. The R^2 value is an indicator of the model fit.

In fact, the model fit is even better when the R^2 value is close to 1, because it indicates that the predictor variables of the model are able to explain most of the total variance of the criterion variable (for example, an equal to 0.40 R^2 means only 40% of the variance in the criterion variable are explained by the predictor variables in the model).

Here, note that the value of a non-standardized regression coefficient (*B*), which reflects the metric of origin of variables, has no limits, and may thus extend from +infinity to -infinity. This is why a high – very high – *B* value does not imply that *X* is a powerful predictor of *Y*. The values of a standardized regression coefficient (β) ranges from – 1.00 to + 1.00 (although they can sometimes slightly exceed these limits). A coefficient β indicates the expected increase of *Y* in standard deviation units while controlling for the other predictors in the model. And as the standard deviation of the standardized variables is equal to 1, it is possible to compare their respective β coefficients.

Regression is called "multiple" when the model has at least two predictor variables. Figures 1.3 and 1.4 show two multiple regression models. It should be noted that the only difference between them lies in the nature, whether of independence, of the predictor variables. They are supposed to be orthogonal (independent or uncorrelated) in the first figure, and oblique (correlated to each other) in the second.

In the first case, the regression coefficient is equivalent to the correlation between the explanatory variable (*X*) and the criterion (*Y*). The second case shows the interest in multiple regression, that of making it possible to distinguish different sources of variance of the criterion variable. Thus, the coefficient obtained here is a regression coefficient that expresses an effect completely independent of all the other effects in the model, giving us information about the correlation between each of the explanatory variables and the criterion as well as about the importance of intercorrelations between all of them.

As can be seen by analyzing the matrix shown in Table 1.1, to estimate figures 1.3 and 1.4 by using the maximum likelihood estimation here – we will discuss this later – the coefficients β on the graph of the first correspond exactly to the correlations between the variables of the model; while those shown in Figure 1.4 are more like partial correlations between these same variables, although they are not correlations. In fact, a coefficient of β 0.54 means that Y could increase by 0.54 deviation for a 1 standard deviation increase of Z, by controlling for X (that is by holding X constant).

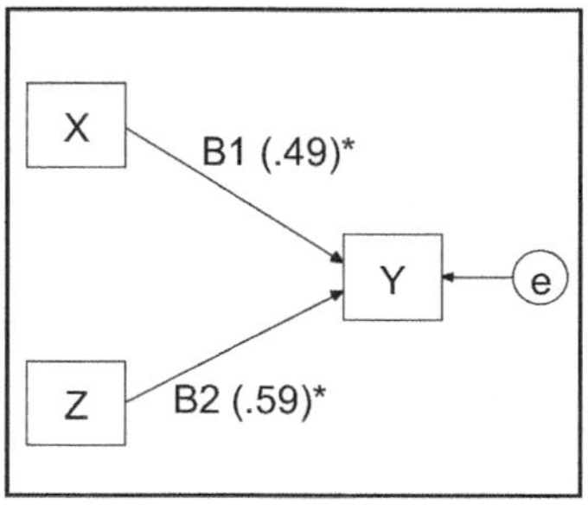

Figure 1.3. *Model of a multiple linear regression (orthogonal predictor/explanatory variables) (*coefficient obtained from [1.1])*

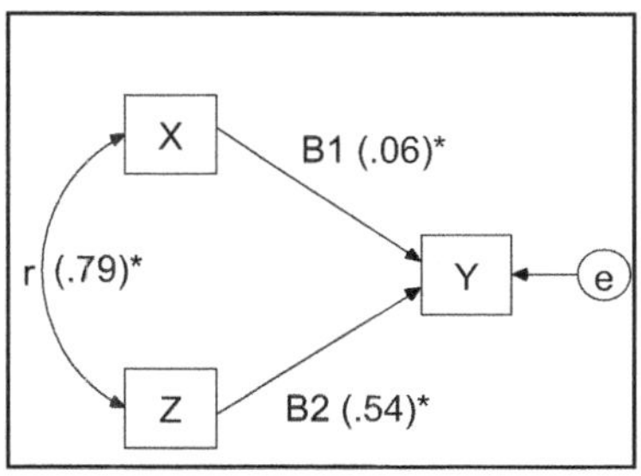

Figure 1.4. *Model of a multiple linear regression (correlated predictor/explanatory variables)*

$$\beta_1 = \frac{r_{YX}-(r_{YZ})(r_{XZ})}{1-r_{XZ}^2} = 0.06 \quad [1.9]$$

$$\beta_2 = \frac{r_{YZ}-(r_{YX})(r_{ZX})}{1-r_{XZ}^2} = 0.54 \quad [1.10]$$

We can thus decompose the variance of Y explained by the regression, that is the square of its multiple correlation (R^2):

$$R^2 = \beta_1 r_{YX} + \beta_2 r_{YZ} \quad [1.11]$$

$$= (0.06)\,(0.49) + (0.54)\,(0.59)$$

$$= 0.34$$

We can easily recognize the specific part attributable to each explanatory variable (*X* and *Z*) through the regression coefficient (β) as well as the common part shared by these same variables through their correlations.

However, it is well known that one of the main difficulties of multiple regression comes from relationships that may exist between the predictor variables. Researchers know nothing about these relationships since the specification of the model consists in simply referring to the criterion and the explanatory variables. If these relationships are more or less perfect, it is either the total confusion of effects or redundancy that affects the results and skews their interpretation (i.e. multicollinearity).

Furthermore, the additive nature of the regression equation seriously limits the possibilities of specifying a model in which relationships between variables can be both direct and indirect. Imagine that a researcher has an idea about the relationships between these explanatory variables. He can reformulate and arrange them by introducing, for example, intermediate explanatory variables in his model that are themselves undergoing the effect of other predictor variables that might have an effect on the criterion. Such an approach requires two conditions: first, a theoretical elaboration for the specified model, and second, an appropriate statistical tool. The approach then becomes confirmatory.

Structural equation modeling, and in this case path analysis, thus provides an answer to this second condition. The problem of multicollinearity, which arises when the correlations between variables are so high that some mathematical operations become impossible, has not been resolved so far. Theoretical conceptualization could, moreover, help researchers not introduce two redundant variables in a path model or make them indicators (for example, items, measured variables) of a latent variable if they opt for a general structural equation model. It is sometimes necessary to get rid of one of the two variables that together show a wide relationship ($r > 0.80$) or have a correlation coefficient higher than the square of the predictor value (R^2) relating to the whole of the explanatory variables. This explains the interest in factor analysis, which helps not only to select the latter (in tracking collinearity), but also to interpret the multiple regressions. For the mathematical aspects, the reader can refer to Gendre [GEN 76]; for multicollinearity in multiple regression, you can consult, among other things, Mason and Perreault [MAS 91] or Morrow-Howell [MOR 94]; and for multicollinearity in SEM, we suggest you to refer to Grewal, Cote and Baumgartner [GRE 04].

1.1.4. *Standard error of the estimate*

The standard error of estimate (SE) is a measure of the prediction error. It evaluates the accuracy of the prediction. This measure will also be used to test the statistical significance of a parameter estimate (B/β). In other words, is the parameter estimate significantly different from zero? Let us take the example of the regression coefficient β equal to 0.49 for our sample ($N = 101$). Is the predictive effect of this coefficient obtained from this sample high enough to be able to conclude with a reasonable probability of risk that it is not zero at the population level from which the sample is taken? To find out, simply divide the regression coefficient (B/β) by its SE. This critical $\left(\frac{B}{SE_B}\right)$ ratio is read as a statistic z whose absolute value, if greater than 1.96, means that $B(\beta)$ is significantly different from zero at $p < 0.05$, whereas an absolute value greater than 2.58 means that $B(\beta)$ is significantly different from zero at $p < 0.01$. In the context of a simple linear regression (Figure 1.2), the SE of the coefficient $B(\beta)$ is obtained as follows:

$$SE_{YX} = \sqrt{\frac{N-1}{N-2}\ var_Y{}^2\,(1 - r_{YX}^2)} \qquad [1.12]$$

where:

- N = sample size;
- var = variance.

Here, we can note the importance of the sample size in calculating the SE. We will address this crucial aspect in SEM later. It should also be noted that SE can serve as an indicator of multicollinearity. In fact, more the predictor variables (IV) are strongly correlated with each other, larger the SE of regression coefficients (thus suggesting hardly accurate predictions), and less likely, therefore, their statistical significance. The reason is simple: more the IV are highly correlated, more it is difficult to determine the variation of the criterion variable for which each IV is responsible. For example, if the predictor variables X and Z, of the model shown in Figure 1.4, are highly correlated (collinear), it is difficult to determine if X is responsible for the variation of Y or if Z is. Therefore, the SE of these variables become very large, their effects on Y inaccurate, and so statistically non-significant. Finally, we can note that the multicollinearity may generate an improper solution, with inadmissible values, such as negative variances.

1.1.5. *Factor analysis*

We purposely keep the factor analysis in the singular – in spite of the diversity of methods that it covers – to keep our discussion general in nature. We will elaborate on some technical aspects that could enlighten us about the benefits of integrating factor analysis into structural equation models later in the book.

The key, to begin with, is in the following note. Fundamentally based on correlations or covariances between a set of variables, factor analysis is about how the latter could be explained by a small number of categories not directly observable. They are commonly called "dimensions", "factors" or even "latent variables", because they are not directly observable.

To set these ideas down, we will start with a simple example. A total of 137 participants were given a scale containing five items. To each item making up the measure, the participant chooses between five possible responses that correspond to a scale ranging from a score of "1" (*completely disagree*) to "5" (*completely agree*) (Table 1.2).

Item/indicator	1	2	3	4	5
In most ways my life is close to my ideal					
The conditions of my life are excellent					
I am satisfied with my life					
So far, I have gotten the important things I want in life					
If I could live my life over, I would change almost nothing					

Table 1.2. *A 5-item scale*

The total scores obtained for all of the items allows, for each individual, the estimation of the construct rated. A high total score is indicative of the presence or the endorsement of the construct measured. This total aggregation offers a comprehensive representation of the model [BAG 94]. It suggests, moreover, a simple model whose nature it is to restore the essence of the measured construct, on condition however that each item is a good indicator. To test this observation, we have the factor analysis. The main application of this method is to reduce the number of intercorrelated variables in order to detect the structure underlying the relationship between these variables. This derived structure, which would be the common source of item variance, is called the "common factor". It is a latent hypothetical variable.

Thus, the covariance (or correlation) matrix constitutes the basic information of the factor analysis. For example, what about the correlations between the five items of our scale, which are considered as measured variables (also called "manifest variables", "indicators" or "observed variables" of our measure (Table 1.2)? Could they be explained by the existence of one or more latent common factor(s)? To know this, this correlation matrix (Table 1.3) is simply subjected to a factor analysis. But first, here is a simple illustration of the reasoning behind the extraction of common factors.

1.1.5.1. *Extraction of common factors*

To simplify our discussion, take the correlations between the first three items of this measure (i.e. measurement tool) scale. Figure 1.5 illustrates the covariations between these items as a diagram. The crosshatched part represents the covariance shared by the three items. The dotted parts represent the covariations between the item pairs. Empty parts represent the variance portion unique to each item.

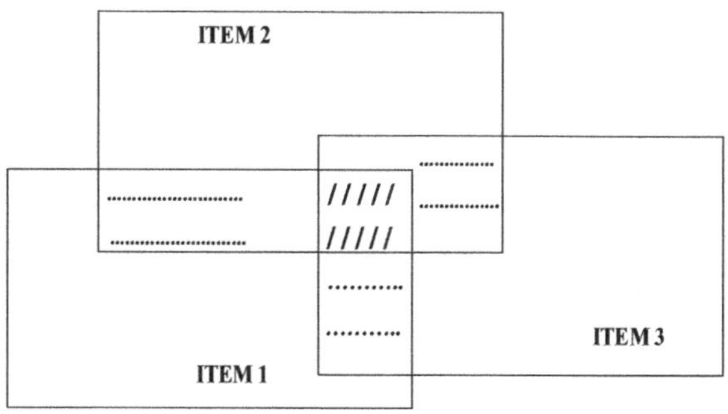

Figure 1.5. *Correlations (covariances) between three items*

Factor extraction procedure allows the identification of the common factor that may explain the crosshatched part, that is what these items may have in common. This factor can partially, or fully, explain the covariances between the three items. Other factors can also be extracted to explain the rest of the covariations outside the crosshatched area. The dotted parts could each be explained by a factor. However, each of these factors will take into account the covariances between two items only and exclude the third. Here, we can note that the common factors extracted do not have precise meaning hitherto. Their existence is, at this stage, purely mathematical. Then comes the interpretation of the results in order to give meaning to and name these factors. Besides, in order to facilitate the interpretation of the extracted factors, we often proceed to factorial rotations, whether they are orthogonal or oblique (the reader eager to know more can consult, among other things, the book by Comrey and Lee [COM 92], or the latest one by Fabrigar and Wegener [FAB 12]).

The passage of the analysis of correlations between variables to the underlying factors requires, in fact, linear combination of these. This combination mathematically takes the form of a regression equation, solving which requires complex estimation procedures. For each factor extracted from the factor analysis, each measured variable (or item) receives a value that varies between – 1 and + 1, called the "factor loading coefficient" (always standardized). The latter defines the importance (i.e. the weight) of the factor for this variable. This means that the correlation between each pair of variables results from their mutual association with the factor(s). Thus, the partial correlation between any pair of variables is assumed to be zero. This local independence allows the searching of factors that account for all the relationships between the variables. In Reuchlin, we find a very detailed and technical presentation on factor analysis methods.

1.1.5.2. *Illustration*

Let us say, for the sake of simplicity, that there is only one factor that would explain the observed correlations between the five items of our measure scale (Table 1.3). Figure 1.6a illustrates this concept because the absence of such a factor means complete independence (orthogonality) of items from each other (Figure 1.6b).

Variable	Item1	Item2	Item3	Item4	Item5
Item1	1.00				
Item2	.37	1.00			
Item3	.57	.30	1.00		
Item4	.26	.31	.39	1.00	
Item5	.34	.33	.43	.43	1.00
SD	.93	.84	.89	.98	1.27

Table 1.3. *Correlations between the five items of the scale shown in Table 1.2 (N = 137)*

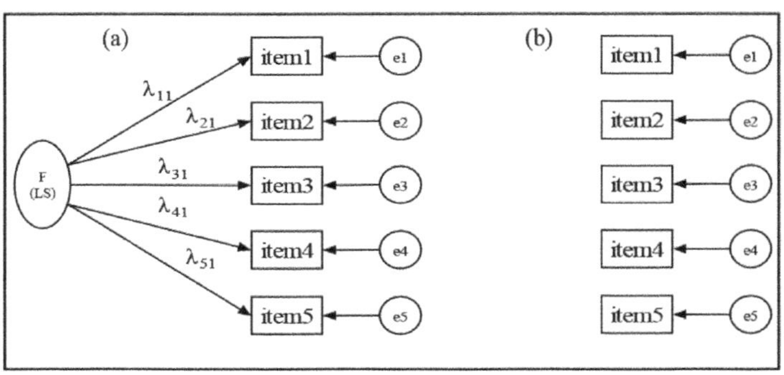

Figure 1.6. *(a) Monofactorial model (one-factor model) of the scale of "life satisfaction (LS)", and (b) null model (also called "independence model" or "baseline model", not to be confused with the null hypothesis)*

Thus, we assume that the variance of responses to each item is explained by two latent factors: one common to all items (factor F), and the other specific to the item (factor e). This specific factor is a kind of combination of item-specific variance and measurement error. We will first seek to estimate the factorial weight of each item, that is the factor loading. In case of an example as simple as ours[3], the factor loading coefficient is none other than the correlation coefficient linking the factor to the measured variable. It also corresponds to the regression coefficient by keeping the factor constant. Thus, there are as many regression equations as there are items:

$$\text{Item 1} = \lambda_{11}F_1 + e_1$$

$$\text{Item 2} = \lambda_{21}F_1 + e_2$$

$$\text{Item 3} = \lambda_{31}F_1 + e_3$$

$$\text{Item 4} = \lambda_{41}F_1 + e_4$$

$$\text{Item 5} = \lambda_{51}F_1 + e_5$$

In other words, each item undergoes the effect of two latent factors: a common factor (F) and a unique factor (e or *U*) referring to a combination of specific factors and the measurement error. It is noteworthy that factor analysis does not provide separate estimates of specific variance and measurement error variance. Lambda "λ" designates the loading (factor weight) of the variables (λ_{11}, λ_{21}, ..., λ_{p1}), represented by the arrow pointing from the common factor to the indicator.

In order to extract factors and solve regression equations, sophisticated parameter estimation methods, such as the principal axis method and the maximum likelihood estimation method are used. It can be noted here that, contrary to the principal axis extraction method, the maximum likelihood method requires a multivariate normal distribution of data, and needs to fix *a priori* the number of factors to be extracted (therefore, it is less exploratory than the principal axis method). Determining the number of factors to be extracted uses certain criteria, such as the Kaiser or Cattell criteria. It should be remembered here that the principal component analysis is not, strictly speaking, factor analysis [FAB 99].

The factor analysis will thus generate a matrix of factor loadings (i.e. model-implied matrix) that best account for the observed correlations between items. Table 1.4 summarizes the factor loadings according to the factor extraction method used. It should be noted that these coefficients shall be interpreted as standardized

3 That is in the case of a theoretical model that has no intermediary variable.

coefficients (β). When squared, a factor loading coefficient of an item makes it possible to give the portion of its variance attributable to the factor on which it depends (underlying factor). It is known as the communality (h^2) of a variable/item. For example, a factor loading of 0.63 squared ($0.63^2 = 0.39$) means 39% of variance explained by the only common factor in our example[4]. The rest of the variance ($1 - 0.39 = 0.61$), that is 61%, is attributable to the specific factor to the item (e, sometimes designated by U^2 for uniqueness). So, even if there is no golden rule, a factor loading coefficient ≥ 0.40 is considered necessary to judge the quality (validity) of an item as an indicator of one factor or another. Some may think that this criterion is quite lax since a factor loading of 0.40 means that only 16% of the explained variance depends on the factor of which one item is the indicator (see the recommendations of [COM 92], or those of [TAB 07] on this subject).

A careful reading of the five items on the scale suggests that this common factor refers to the "life satisfaction" construct. It is in fact the "life satisfaction" scale of Diener, Emmons, Larsen, Griffin [DIE 85] and [GRI 85].

	Factor loadings	
Variable	Principal axis	Maximum likelihood
Item1	.62	.63
Item2	.52	.51
Item3	.70	.71
Item4	.56	.56
Item5	.63	.63

Table 1.4. *Factor loadings for 5 items depending on factor extraction method*

Remember also that these are factor loading coefficients that allow the reproducing of the correlation matrix (Σ) (model-implied matrix). In fact, in the case

4 In the presence of several extracted factors, the commonality of an item is the sum of its factor loading squared by each factor. Thus, h^2 here refers to the portion of variance attributable to all the factors generated by the factor analysis.

of a monofactorial solution[5], the product of two factor loadings allows reproducing their correlation. Thus, model-based reproduced correlation between item 1 and item 2 is equal to (0.62) (0.52) = 0.32 using the principal axis method, and (0.63) (0.51) = 0.32 using the maximum likelihood estimation method. From the observed correlations, the reproduced correlations are deduced term by term. Differences, whether positive or negative, form the residual correlation matrix. Table 1.5 shows the three matrices involved: original (S), reproduced (Σ), and residual (S – Σ). The last two matrices are obtained using the maximum likelihood estimation method. The same rule applies to other methods. We will come back to this in Chapter 3 of this book, when we will deal with confirmatory factor analysis.

	Reproduced matrix (Σ)					**Residual matrix (S - Σ)**				
	Item 1	Item 2	Item 3	Item 4	Item 5	Item 1	Item 2	Item 3	Item 4	Item 5
Item 1	_	*3.2*	*0.45*	*0.35*	*0.39*	_				
Item 2	0.37	_	*0.36*	*0.28*	*0.32*	**0.05**	_			
Item 3	0.57	0.30	_	*0.39*	*0.44*	**0.12**	**– 0.06**	_		
Item 4	0.26	0.31	0.39	_	*0.35*	**– 0.09**	**0.03**	**0.00**	_	
Item 5	0.34	0.33	0.43	0.43	_	**– 0.05**	**0.01**	**– 0.01**	**0.08**	_
	Observed matrix (S)									

Table 1.5. *Correlation matrices (reproduced above diagonal, observed below diagonal, as well as residual) of items of the life satisfaction scale; maximum likelihood estimation method*

5 Things are obviously more complicated with a multifactorial solution where the correlation reproduced between X_1 and $X_2 = \Sigma_m \lambda_{X1m} \lambda_{X2m}$ (m denotes the number of factors).

Mostly exploratory and descriptive[6], factor analysis has at least two advantages. First, direct use making it possible to: (1) identify groups of correlated variables in order to detect the underlying structures – the aim of this method is to determine the organization of variables, by identifying those that are similar and those that contrast with each other; (2) to reduce the set of variables into a smaller number of factors. In fact, it is reasonable to think that the variables that display a high degree of similarity are assumed to measure the same construct, the latter being a non-observable theoretical entity called "factor". It is expected to account for the covariations between the measured variables by clearing the "belief", or the "perception", that underlies them. The "life satisfaction" construct is one such example. It is a theoretical construction that is assumed to be evaluated by indicators, or measured variables, which are the items. The perception that each participant has life satisfaction is estimated by his/her way of answering the items, for which correlations are used as an index. Associations – of high degrees – between items can be explained by the fact that they involve the same perception, or the same core belief. Second, an indirect use helps to process data – preparing them – to simplify them and avoid artifacts, such as multicollinearity (reflecting a redundancy of variables), that make a matrix ill-conditioned.

1.1.6. *Data distribution normality*

The normality of data distribution is at the heart of structural equation modeling because it belongs to linear models. Moreover, LISREL, still used as the generic name of this technique (specifically, we still talk of a "Lisrel model" to denote structural models), is in fact an abbreviation of LInear Structrural RELationships, the model and first SEM program, designed by Jöreskog [JÖR 73, JÖR 73].

The statistical tests used by researchers to assess these models are, in fact, based on the hypothesis of normal distribution: multivariate normal distribution of data. The entire scope of statistical inference depends on it. In order to determine the level of significance of the tests obtained, we need a theoretical model, which will tell us the probability of an error involved in the acceptance (or rejection) of the relationship between two variables depending on the size of our sample. This model is the so-called mathematical "normal distribution". Indeed, most statistical procedures used in SEM involve the assumption of normality of distribution of observations in their derivation. A "univariate normal distribution" is when a single variable follows the

6 The confirmatory nature is hardly absent from factor analysis, as Reuchlin (1964) so aptly described, especially with the maximum likelihood estimation method, which is accompanied by the statistical test for the fit of the factor solution.

normal distribution, and in the "multivariate normal distribution", a group of variables follows the normal distribution.

Now, some simple questions are raised: what is a normal distribution? How can this normality be assessed? What to do in case of violation of this assumption?

Curved and bell shaped, the normal distribution is defined by two parameters: the mean, center of the distribution setting the symmetry of the curve, and deviation, whose value affects the appearance of the latter[7]. It is a unimodal and symmetric distribution. Standard normal distribution with a mean of 0.00 and standard deviation of 1.00 is its most convenient and simplest form. Such a transformation does not affect the relationship between values, and its result is the z-score (see [1.3]).

We can see from a normal curve that a distribution is said to be "normal" when 95.44% of its observations fall within an interval which corresponds to, and in absolute value, twice (or more precisely 1.96 times) the standard deviation from the same mean. These are limits within which 95.44% observations are found. As for the 4.6% of the observations (2.3% at each end of the curve), they obtain higher z-values in absolute value at 2.00 (to be more precise, at 1.96) times the standard deviation from the mean. The z-score is of crucial importance in SEM as we will use it, through the critical ratio, to test the hypothesis of statistical significance of model parameter estimates. We have already seen that an estimated parameter is considered significant when it is statistically different from zero. We will thus assess its utility and its pertinence in the model. It is important to keep in mind that some estimation methods require the hypothesis of multivariate normality. Unbiasedeness, effeciency as well as precision of the parameter estimates depend on it.

A normal empirical distribution is supposed to be superimposed on such a theoretical distribution. But there could be questions about the isomorphism of such a distribution with reality, about the reasons that make us believe that this distribution is a mathematical translation of empirical observations. In reality, there is no distribution that fully complies with the normal curve. This way of working may seem reasonable for researchers in some cases, and quite questionable in others. But, as nature does not hate irregular distributions in the famous words of Thorndike [THO 13], there are many phenomena that differ from this mathematical model. However, it is important to specify that the distribution of the normal curve constitutes a convenient model that gives a technical benchmark following which the empirical results may be assessed.

7 In a normal distribution, the mode and median are the mean.

There are several indices that allow for statistically estimating the normal distribution of the curve. Among the best known, we name two that provide very useful indications in this respect: kurtosis concerning the more or less pointed appearance – or shape – of the curve, and skewness (asymmetry) that concerns the deviation from the symmetry of the normal distribution. A value significantly different from zero (with a 5% risk of error, particularly higher than 3 in absolute value) indicates that the form or the symmetry of the curve has not been respected. Regarding Kurtosis, a positive value signifies, in this case, that the distribution curve is sharper (called "leptokurtic"). In the case of a measurement tool (e.g. test), it means that there are neither very easy items nor very hard items, but too many items of average difficulty. On the other hand, a negative value suggests that the distribution has a flat shape (platykurtic) that indicates, in case of a test, that there are too many very easy items (ceiling effect) and too many very difficult items (floor effect). In the case of skewness, when the index is positive, the grouping of values is on the left side (too many difficult items). If the index is negative, the grouping is on the right side (too many easy items). Thus, a peak with an absolute value higher than 10 indicates a problematic univariate distribution, while a value higher than 20 indicates a significant violation of this distribution (the reader will find illustrated presentations of asymmetry and kurtosis in [BRO 16]).

Here, the question raised is about knowing whether the value of indices is sufficiently high to be able to conclude, with reasonable probability, that there is a risk of violating the form of the curve. To answer it, it is enough to divide the value of the index by its standard error (*SE*). This critical ratio reads like a *z*-test, of which an absolute value higher than 1.96 means that the index (asymmetry or kurtosis) is significantly different from zero to $p < 0.05$, while an absolute value higher than 2.58 means that it is significantly different from zero to $p < 0.01$. In both cases, it can be concluded that there is a violation of form of data distribution.

But the review of the univariate skewness and kurtosis indices no longer seems sufficient to prejudge multivariate distribution. The use of multivariate normality tests, like the Mardia [MAR 70] coefficient, now proves necessary and very practical. Let us note that a high value at the latter, especially higher than 5 in absolute value, indicates a deviation with respect to multivariate normality. This coefficient is available in almost all structural equation modeling software; it is often accompanied by its *z*-value, thus making it possible to judge its statistical significance. These SEM software have made possible extreme outlier detection, which might be the cause of deviations from normality. This makes it possible to identify observations that contribute greatly to the value of the Mardia coefficient. But, it is sometimes enough to eliminate them to return the normality to the distribution [KOR 14].

But, in the presence of data seriously violating normality, some transformations that normalize them could be a remedy. Yet, even though these transformations, in this case normalization, are processes that have been well known for long and even have some advantages, it is rare to see research in SEM using them (see [MOO 93]). Also, for ordinal and categorical scales, the aggregation of items (indicator), especially when there are multiple, may sometimes be a remedy to the non-normality of distributions mostly because they are many in number. In this case, the analysis no longer concerns all the items constituting the measure (total disaggregation), but the item parcels referring to subsets of items whose scores have been summed [LIT 02]. This is a total or partial aggregation of indicators/items of a measure. The reader will find details of use as well as examples of application of this procedure suggested by Bernstein and Teng [BER 89] in the works of Bagozzi and his collaborators [BAG 94, BAG 98, ABE 96]. However, creating item parcelling requires at least two conditions: the measure has to be one-dimensional (i.e. a congeneric model) and items constituting each parcel created have to be randomly selected. For limits concerning item parcelling, the reader can refer to [MAR 13].

However, it should be specified that when the treatment of non-normality is not conclusive (despite item parcelling), and the deviation with respect to normality remains high, it is necessary to consider using the most appropriate statistical indicators and estimation methods. We will discuss more on that later.

1.2. Basic principles of SEM

Similar to factor analysis, the reproduced matrix is a central element in structural equations modeling. To simplify, the crux is in the following clarification: the starting point of structural equation modeling also involves comparing the covariance (or correlation) matrix of the research data (S) with a covariance matrix that is reproduced (Σ) from the theoretical model that the researcher wishes to test (*model-implied covariance matrix*). The theoretical model specified by the represents the null hypothesis (H_0) as a model assumed plausible. The purpose of this comparison is to assess the fit between the observed variables and the selected conceptual model. If the reproduced covariance matrix is equal to the observed matrix ($\Sigma = S$), it refers to the model's fit to the data, or fit between the model tested and observed data. In other words, the null hypothesis ($H_0 : \Sigma = S$) is not rejected and the specified model is acceptable.

Differences between the two matrices are represented in the residual covariance (or correlation) matrix, which is obtained by subtracting the value of each element of the reproduced matrix from the corresponding element in the observed matrix

(S – Σ). The degree of similarity between the two matrices indicates the overall degree of the model's fit to the data.

The covariance matrix (Σ) is reproduced using the model parameter estimates. Estimating involves finding a value for each unknown parameter in the specified hypothetical model. But, it goes well beyond this simple objective, as it is important that the calculated values allow for reproducing a covariance matrix (Σ) that resembles the observed covariance matrix (S) as much as possible. It is an iterative procedure whose general principle is to start from an initial value (specified for the set or for each individual parameter either by the user or automatically by the software) and to refine it progressively by successive iterations that stop when no new value for each parameter makes it possible any longer to reduce the difference between the observed covariance matrix and the reconstituted covariance matrix. These different operations are performed by algorithms for minimization (discrepancy functions or minimization functions, F_{min}) that, despite having the same purpose of finding the best parameter estimates, nevertheless differ from the mathematical function used for it, that is for minimizing the discrepancy between the observed covariance matrix and the reproduced covariance matrix.

In order to illustrate this point, we will go back to the matrix in Table 1.1, and propose another model for it. Shown in Figure 1.7a, this model has the advantage of being simple: it has two parameters to be estimated, namely P_1 and P_2. These are two regression coefficients *B* (β). It is thus easy to test the iterative procedure by applying the ordinary least squares (OLS) method, whose purpose is to minimize a particular discrepancy function that is defined as the sum of the squares of the differences between the observed correlations and the reproduced correlations (Σd^2). Table 1.6 summarizes the whole iterative method estimation.

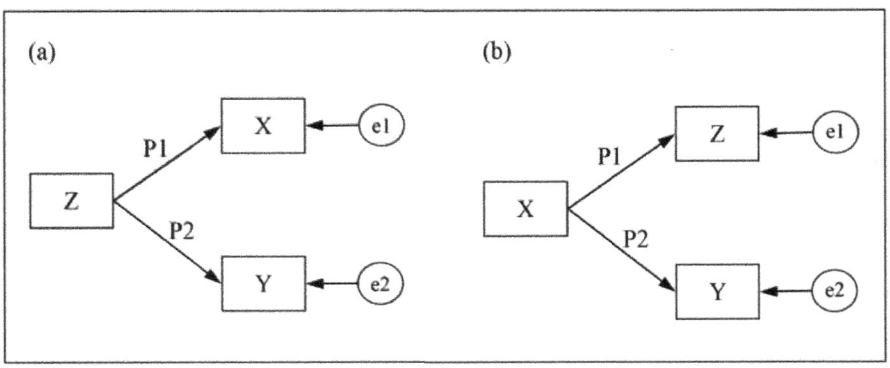

Figure 1.7. *Equivalent path models (a) and (b) linking three observed variables*

To begin, all estimation methods require that the starting values for all parameters to be estimated be specified. Some programs also allow for determining the maximum number of iterations allowed, and specifying a convergence condition that, once met, causes the iterative procedure to stop.

To carry out the first iteration, we have arbitrarily assigned a value of 0.50 to each of the two parameters P_1 and P_2 of the model. These values allow us to first reproduce a correlation matrix: $r_{XZ} = 0.50$, $r_{YZ} = 0.50$, and $r_{XY} = 0.25$. If we now apply the OLS method, we then get the next discrepancy function, $F_{min} = (0.79 - 0.50)^2 + (0.59 - 0.50)^2 + (0.49 - 0.25)^2 = 0.149$ (Table 1.6).

The smaller the value of this function, the better the fit between the observed covariance matrix and the reproduced covariance matrix. Moreover, this value becomes equal to zero when parameter estimations allow for perfectly reproducing the observed matrix.

Thus, the initial values will be systematically changed from one iteration to another up to the moment when the iterative process will end, that is when no new value is able to improve the discrepancy function.

For example, changes introduced in steps 1a, 1b, and 1c of Table 1.6 aim to determine the effect that they can have on the discrepancy function (F_{min}). It is shown that the simultaneous reduction of the P_1 and P_2 values deteriorates this function (see cycle 1a with respect to cycle 1). An alternative fall of these values confirms this tendency (see cycles 1b and 1c with respect to cycle 1). Subsequent steps aim to reverse the first trend. It remains to be determined how much the P_1 and P_2 can increase. It is clear that at this level, a P_2 value higher than 0.61 causes an impediment to the minimization function (see cycle 3a). On the other hand, progressive increase of P_1 improves this function. Finally, it can be noted that it is better to stop the process at step 4c, as the last step (i.e. step 5) begins deteriorating minimization, which goes from 0.0006 to 0.0008. And it is by obtaining 0.80 and 0.61 respectively that P_1 and P_2 allow for the best minimization. The iteration process can then stop with a discrepancy function with a of value $F_{min} = 0.0006$.

The remaining differences between the observed correlations and the reproduced correlations on the basis of the estimated parameters represent elements of the residual matrix. Table 1.7 provides reproduced and residual correlation matrices of the model in Figure 1.7a, obtained by the OLS method.

			Observed correlations			**Discrepancy function (F_{min})**
	Parameter values		$r_{XZ} = 0.79$	$r_{YZ} = 0.59$	$r_{XY} = 0.49$	$\sum d^2$
Iteration cycles	**P1**	**P2**	**Reproduced correlations**			**Least squares**
1	0.50	0.50	0.50	0.50	0.250	0.149
1a	0.49	0.49	0.49	0.49	0.240	0.162
1b	0.49	0.50	0.49	0.50	0.245	0.158
1c	0.50	0.49	0.50	0.49	0.245	0.158
2	0.55	0.55	0.55	0.55	0.300	0.094
2a	0.60	0.60	0.60	0.60	0.360	0.029
3	0.65	0.61	0.65	0.61	0.400	0.027
3a	0.65	0.62	0.65	0.62	0.403	0.028
4	0.67	0.61	0.67	0.61	0.408	0.021
4a	0.70	0.61	0.70	0.61	0.427	0.012
4b	0.75	0.61	0.75	0.61	0.457	0.003
4c	**0.80**	**0.61**	**0.80**	**0.61**	**0.480**	**0.0006**
5	0.81	0.61	0.81	0.61	0.494	0.0008

Table 1.6. *Solution of the iterative procedure for the model in Figure 1.7a*

	Reproduced correlations (Σ)			**Residual correlations (S – Σ)**		
	X	*Y*	*Z*	*X*	*Y*	*Z*
X	_	*0.48*	*0.80*	_		
Y	0.49	_	*0.61*	**0.01**	_	
Z	0.79	0.59	_	**– 0.01**	**– 0.02**	_
	Observed correlations (S)					

Table 1.7. *Original, reproduced, and residual correlation matrices of the model in Figure 1.5a, using the OLS estimation method*

It can be noted that the smallness of the discrepancy between the observed and reproduced correlations of three variables (0.01, – 0.01, and – 0.02) proves the similarity between the two matrices and consequently, the plausibility of the model. Approximation – which can be assessed based on goodness-of-fit indices – seem sufficient at first sight to even talk about the fit of the model, in other words the adequacy of the theory on which the facts are based. To be convinced of this, let us evaluate an alternative model in which the variable X (chocolate consumption) influences variables Z (milieu) and Y (life satisfaction). Figure 1.7b shows this model and Table 1.8 summarizes the iterative process generated by the OLS estimation method.

				Observed correlations		**Discrepancy function (F_{min})**
	Parameter values		$r_{XZ} = 0.79$	$r_{YZ} = 0.59$	$r_{XY} = 0.49$	$\sum d^2$
Iteration cycles	**P1**	**P2**		**Reproduced correlations**		**Least squares**
1	0.50	0.50	0.50	0.50	0.250	0.1998
1a	0.49	0.49	0.49	0.49	0.240	0.2125
1b	0.51	0.51	0.51	0.51	0.260	0.1877
2	0.52	0.52	0.52	0.52	0.270	0.1762
2a	0.58	0.58	0.58	0.58	0.336	0.1167
2b	0.60	0.60	0.60	0.60	0.360	0.1011
2c	0.61	0.61	0.61	0.61	0.372	0.0943
2d	0.62	0.62	0.62	0.62	0.384	0.0882
2e	0.63	0.63	0.63	0.63	0.397	0.0824
2f	0.64	0.64	0.64	0.64	0.409	0.0777
2g	0.70	0.70	0.70	0.70	0.49	0.0622
2h	0.80	0.80	0.80	0.80	0.640	0.0996
3	0.80	0.70	0.80	0.70	0.560	0.0451
3a	0.85	0.65	0.85	0.65	0.552	0.0328
3b	0.88	0.60	0.88	0.60	0.528	0.0240
4	0.89	0.59	0.89	0.59	0.525	0.0242
4a	0.88	0.59	0.88	0.59	0.519	0.0231
4b	0.88	0.58	0.88	0.58	0.510	0.0226
4c	**0.88**	**0.57**	**0.88**	**0.57**	**0.501**	**0.0224**
4d	0.88	0.56	0.88	0.56	0.492	0.0226

Table 1.8. *Solution of the iterative procedure for the model in Figure 1.7b*

We remember that the iterative process must end when no new value is able to improve the discrepancy function any longer. It should be noted in Table 1.8 that after a cycle of improvement of the discrepancy function, the last step 4d spells a reversal of the situation. In this case, the estimation procedure should be stopped and the result from the previous step 4c should be kept. The discrepancy function thus gets a value equal to 0.0224, and parameters P_1 and P_2 get 0.88 and 0.57 respectively.

	***Reproduced correlations* (Σ)**			**Residual correlations (S – Σ)**		
	X	*Y*	*Z*	*X*	*Y*	*Z*
X	_	*0.57*	*0.88*	_		
Y	0.49	_	*0.50*	**– 0.08**	_	
Z	0.79	0.59	_	**– 0.09**	**– 0.09**	_
	Observed correlations (S)					

Table 1.9. *Original, reproduced, and residual correlation matrices of the model in Figure 1.5b, using the OLS estimation method*

It is also remembered that the residual matrix provides clues as to whether the specified model is able to adequately reproduce the original correlation matrix (or variance-covariance matrix). In fact, it makes it possible to know the degree of approximation of the observed matrix, the degree of similarity between the latter and the reproduced matrix based on the model that we intend to use to describe original correlations. It can be noted in Table 1.9 that the discrepancy between the observed matrix and the reproduced matrix is such that we are right in thinking it is a model totally inconsistent with the data. We will discuss more on that later.

We have already underlined the fact that, apart from the experimental method, all other methods seem unfit to determine a causal link in a strict manner. It is true that nothing in a correlation matrix allows for changing the relational nature between the variables that it takes into account into a causal nature between them. However, while the procedure used to test the theoretical assumptions formalized by figures 1.7a and 1.7b has actually failed to demonstrate causal links, it showed that these links could now not be read equally in one way or the other (see Chapter 5 of this book). It will merely be noted that fit with the facts of a model for the benefit of another has highlighted a certain orientation in different connections. It is not absurd to think that, despite their high correlation, it is variable Z (in this case, the "milieu") that has an effect on variable X (in this case, "chocolate consumption"), and not the other way around.

Moreover, the procedure that follows and leads to comparing matrices can be surprising. In fact, one observed correlation matrix is compared to a matrix that is derived from path coefficients that are themselves based on an estimation from this first observed matrix. Sometimes, the correlation coefficient easily reproduces the regression coefficient. It is a mathematical tautology that guarantees a perfect prediction [JAC 96]. Such a mathematical tautology that renders any comparison useless is the prerogative of saturated models (just-identified) that we will discuss later.

In Chapter 3 of this book, we will see how to reproduce a covariance matrix (Σ) from the parameters of a simple measurement model (see the topic of confirmatory factor analysis in the chapter). In what concerns general structural equation models – including latent variables – the derivation of a covariance matrix from the parameters of the estimated model is obviously more complicated because of the simultaneous presence of the measurement model and the structural model. The reader eager to know the details can consult the work of Mueller [MUE 96] among others.

1.2.1. *Estimation methods (estimators)*

As we just saw, the model estimation involves finding a value for each unknown (free) parameter in the specified model. But, it goes well beyond this simple objective, as it is important that the calculated values allow reproducing a covariance matrix (Σ) that resembles the observed covariance matrix (S) as much as possible. It is an iterative procedure whose general principle is to start from an initial value (specified for the set or for each individual parameter either by the user or automatically by the software) and to refine it progressively by successive iterations that stop when no new value for each parameter makes it possible any longer to reduce the difference between the observed covariance matrix and the reproduced covariance matrix. These different operations are performed by minimization algorithms (i.e., discrepancy or minimization function) that, despite having the same purpose of finding the best parameter estimates, nevertheless differ from the mathematical function used for it, that is for minimizing the deviation between the observed covariance matrix and the reproduced covariance matrix. This often involves complex mathematical functions based on matrix algebra (vectors, inverse of a matrix, weighted matrix, determinant, etc.) of a level higher than that desired for this introduction, but whose details could be found in any specialized book (for example, see [BOL 89]). Thus, so much more than their purely mathematical aspects, what interests us here, in the point of view of the user wanting to get unbiased estimates of his/her model's parameters, is to know the considerations that can guide the choice of an estimation method. In fact, since sample parameter estimates are used to infer population parameter estimates, the first must be, among other things, unbiased, accurate, and consistent. And estimator is crucial here, hence the interest that it can be the object of a deliberate and justified choice, which alone will make it possible to retain that which is most appropriate to the data present [LAD 89]. *Grosso modo*, this choice boils down to two options dictated by the type of data and, in particular, by the nature of their distribution. The first concerns estimation methods that require the hypothesis of multivariate normality of data, while the second concerns estimators that are most suitable to data that deviate following the normal distribution. The specificities, advantages, and disadvantages of these estimators have given rise to fairly abundant and rich work and publications [CHO 95, WES 95].

These methods all have the same main objective of rendering, iteratively, discrepancy (F_{min}) as tiny as possible between two matrices. The value of the function is positive, even equal to zero when $S = \Sigma$. This means, in this case, that the model is perfectly compatible to the data, in other words, $H_0 : F_{min} = 0.00$.

The major difference between these methods lies in the manner in which the mathematical discrepancy function ($F[S, \Sigma] = F_{min}$) is used to minimize deviations between the observed correlation matrix (for example, Pearson correlations, polychoric correlations, tetrachoric correlations) and the reproduced correlation matrix. Once this objective has been met, the statistical significance remains to be assessed. To this end, we use the χ^2 test that is calculated in the following way in lavaan:

$$\chi^2 = (N)F_{min} \qquad [1.13]$$

where:

– N is the sample size;

– F_{min} denotes minimum discrepancy ($F[S, \Sigma]$) obtained by the estimation method used (for example, F_{ML}, F_{GLS}, F_{WLS}, which we will discuss later).

This statistical test allows for judging whether the null hypothesis (H_0: $S = \Sigma$) is admissible, namely that there is no significant difference between the two matrices. The χ^2 value, which tends to increase with the F value, is all the greater because the two matrices compared are dissimilar from one another. A significant (high enough) χ^2 makes it possible to reject the null hypothesis, thus indicating that the specified model does not allow for adequately reproducing the observed correlation matrix. However, when χ^2 is equal to zero (or insignificant), namely when the discrepancy is zero (F[S, Σ] = 0.00), it means that there is a perfect (or near perfect) fit between two matrices. Reading a statistical table of distribution of this index is based on the degrees of freedom (*df*). These are obtained by subtracting the number of parameters to be estimated in the model (t) from the number of variances and covariances available, that is to say k(k + 1)/2 where k denotes the number of observed variables:

$$df = [k\,(k + 1)/2] - t \qquad [1.14]$$

The advantage of χ^2, which is the only statistical test in SEM, is that it allows for proving the statistical significance of the model's fit to the data, under certain conditions concerning, in particular, the nature of data distribution and the sample size.

Limits around this test are multiple and now, well known. Apart from its sensitivity to sample size (the bigger it is and higher the risk of the model being rejected[8]), the multivariate normal data assumption, which is required for this test. It is true that in humanities and social sciences, data that perfectly respects normality is rarely available.

It should also be noted that the sample size directly affects the χ^2 value. The sensitivity of this index to the sample size has raised some well-founded reservations that have led to the emergence of other complementary goodness-of-fit indices that will be discussed later. As for the choice of one estimation method over another, it is an important aspect and we will discuss it later.

1.2.1.1. *Estimators for normally distributed data*

There are two estimators for normally distributed data that are commonly used. They are the maximum likelihood method (ML, F_{ML}) and the Generalized Least Squares (GLS, F_{GLS}).

With the maximum likelihood method (ML), the discrepancy function (F_{min}) takes the following formula:

$$F_{ML} = \log|S| - \log|\Sigma| + tr(S\Sigma^{-1}) - k \qquad [1.15]$$

8 lavaan provides an index called Hoelter's critical N indicating the maximum sample size required to accept a model at a given level of probability of χ^2 (0.05 or 0.01) (see [HU 95], for the formula of this index).

where:

– refers to the natural logarithm function (base *e*);

– || is the determinant of the matrix;

– k is the number of variables in the correlation (or covariance) matrix;

– is the trace matrix algebra function which sums diagonal elements;

– S = observed matrix;

– Σ = reproduced matrix;

– Σ^{-1} = inverse of matrix Σ.

The function used by the Generalized Least Squares (GLS) estimation method is written as follows:

$$F_{GLS} = 1/2 \times tr[S^{-1}(S - \Sigma)]^2 \qquad [1.16]$$

where:

– tr = trace of the matrix (or the sum of the diagonal elements of the matrix);

– S = observed matrix;

– Σ= reproduced matrix;

– S^{-1} = inverse of matrix S.

Finally, the third known estimation method, the Weighted Least Squares method (WLS, see [BRO 84]), based on the polychoric correlation matrix, is not recommended for samples of too small a size. However, unlike the previous ones, this method has the advantage of not depending on the form of data distribution.

1.2.1.2. *Which estimators for non-normally distributed data?*

Let us recall that the equivalent of the aforementioned method is also known as the Asymptotic Distribution-Free function (F_{ADF}), and as the Arbitrary Generalized Least Squares (F_{AGLS}) function (i.e. estimator). Much later, we will discuss other estimation methods, some of which seem to be more appropriate for ordinal/ categorical variables and for data whose distribution deviates from normality.

The debate on the performance of estimators based on the type of data to be analyzed (i.e. continuous variables, ordinal variables, binaries, normality of distribution) is still open [LI 16]. In fact, although it is the default estimator in all modeling software (including lavaan), the maximum likelihood estimation method (ML) which requires that variables are continuous and multivariate normal. In humanities and social sciences, there are at least two challenges that are faced in using this estimator. First, the prevalence of ordinal (for example, Likert-type scale) and dichotomous/binary (true/false) outcome measures (indicators). Second, the prevalence of non-normally distributed data. From a purist point of view, the maximum likelihood method is not at all appropriate for ordinal measurements such as Likert scale items. In fact, it is now known that such measurements often display a distribution that deviates from normality [MIC 89].

Several simulations have shown that the application of the maximum likelihood estimation method (or generalized least squares) on data that does not have a normal distribution also affects the estimation of standard errors of the parameters than the goodness-of-fit statistics: some fit indices are overestimated, and the χ^2 tends to increase as the data gap increases with respect to normality [WES 95]. However, the findings of Chou and Bentler [CHO 95] make it possible to qualify these remarks. These authors showed that in the presence of a sufficiently large sample, maximum likelihood estimation method and generalized least squares method do not make the results suffer, even when the multivariate normality is slightly violated (see also [BEN 96]). The robustness of these methods is not always guaranteed. In case of a more serious violation, several options are available to SEM users [FIN 13]. They can be classified into three categories.

The first groups the family of the maximum likelihood estimation method with corrections of normality violations (Robust ML). It concerns new estimation methods considered to be more "robust", as statistics (i.e. standard errors and χ^2) that they generate are assumed to be reliable, even when distributional assumptions are violated.

The Satorra-Bentler χ^2 (SBχ^2) incorporates a scaling correction for the χ^2 (called scaled χ^2). Its equivalent in lavaan is the MLM estimator.

It is obtained as follows:

$$\mathrm{SB}\chi^2 = \frac{\mathrm{ML}\chi^2}{d} \quad [1.17]$$

where:

– d = the correction factor according to the degree of violation of normality;

– ML = maximum likelihood estimator.

In fact, in the absence of any violation of the multivariate normality, $SB\chi^2 = ML\chi^2$. The value of this correction factor (*d*) is given in the results under the name scaling correction factor for the Satorra-Bentler correction.

The Yuan-Bentler χ^2 ($YB\chi^2$) is a robust ML estimator similar to the aforementioned one, but more suited for a small sample size. Its equivalent in lavaan is the MLR estimator. The value of the correction factor (*d*) is provided in results under the name scaling correction factor for the Yuan-Bentler correction, allowing for calculating the $YB\chi^2$ as follows:

$$YB\chi^2 = \frac{ML\chi^2}{d} \qquad [1.18]$$

The second category groups alternative estimation methods to ML:

a) first, the weighted least squares method (WLS). This estimator, which analyzes the polychoric or polyserial correlation matrix has at least two disadvantages. It requires a fairly large sample size [FLO 04] to hopefully get stable results (for example, at least 1,200 participants for 28 variables, according to [JÖR 89], pp. 2–8, Prelis). And, above all, it quite often runs into convergence problems and produces improper solutions where complex models are estimated. A negative variance, known as the "Heywood case", makes the solution improper because, as we remember, a variance can hardly be negative;

b) the Diagonally Weighted Least Squares (DWLS) method is next. Jöreskog and Sörbom [JÖR 89] encouraged using this method when the sample is small and data violates normality. This estimator, for which the polychoric or polyserial correlation matrix serves as the basis for analysis, is a compromise between the unweighted least squares method and the full weighted least squares method [JÖR 89]. Two "robust" versions of DWLS that are close to this estimator, called "WLSM" and "WLSMV" in lavaan (and M*plus*), give corrected estimates improving the solution outcomes (standard errors, χ^2, fit indices described as "robust").

These methods use a particular calculation of the weighted matrix as a basis and are based on the generalized least squares method. The estimation procedure that requires the inversion of the weighted matrix generates calculations that become problematic when the number of variables exceeds 20, and require a large sample of participants in order to have stable and accurate estimates. Another limitation is the requirement to analyze raw data, and therefore have it.

The third method refers to the resampling procedure (bootstrap). This procedure requires neither a normal multivariate distribution nor a large sample (but it is not recommended for dichotomous and ordinal measures with few response categories). MacKinnon, Lockwood, and Williams [MAC 04] believe that it produces results (for example, standard errors) that are very accurate and reliable.

The principle of this procedure is simple [MAC 08]. A certain number of samples (set by the researcher, e.g. "N bootstrap = 1000") are generated randomly with replacement from the initial sample considered, as the population. Each generated sample contains the same number of observations as the initial sample. Then there will be as many estimates of the model parameters as there are samples generated. An average of estimations of each model parameter is calculated, with a confidence interval (CI). An estimate is significant at $p < 0.05$ if its confidence interval at 95% (95% CI) does not include a null value (see [PRE 08a]). The resampling procedure, with lavaan, also has the possibility of getting confidence intervals of fit indices.

Table 1.10 summarizes the recommendations concerning estimators available in lavaan based on the type of data to be analyzed.

Data type and normality assumption	Recommended estimator
Continuous data	
1- Approximately normal distribution	ML
2- Violation of normality assumption	ML (in case of moderate violation) MLM, MLR, Bootstrap
Ordinal/categorical data	
1- Approximately normal distribution	ML (if at least 6 response categories) MLM, MLR (if at least 4 response categories) WLSMV(binary response or 3 response categories)
2- Violation of normality assumption	ML (if at least 6 response categories) MLM, MLR (if at least 4 response categories) WLSMV (in case of severe violation)

Table 1.10. *Recommendations concerning the main estimators available in lavaan according to the type of data*

Eventually, the choice of estimation method depends on four criteria:

– first, the measurement level of data: it seems well established that the most appropriate estimator for binary/dichotomous variables is WLS [MUT 93] and its recent extension (WLSMV). Following the work of Muthén [MUT 83, MUT 84], it quickly became aware of the need to change the approach to the data obtained with binary or ordinal scales;

– data distribution properties, as we saw;

– available data: raw data or correlation/covariance matrix? Although the covariance matrix is the basis of any structural analysis, except the ML method, all other methods require using raw data. In the absence of raw data, one can instead use either a correlation matrix or a variance-covariance matrix;

– finally, sample size. This last point deserves attention because it is linked with statistical power.

1.2.1.3. *Sample size and statistical power*

By opting for SEM, the researcher must immediately look at the crucial and throbbing question of the necessary number of participants to be collected to hope to obtain a proper solution, an acceptable level of accuracy and statistical power of estimates of his/her model's parameters, as well as reliable goodness-of fit indices. Today, specialists are unanimous in considering that structural equation modeling requires a lot of participants. Its application to sample sizes that are too small may bias the estimates obtained. However, it remains to be seen how many participants are needed and sufficient to obtain accurate and reliable estimates. Several general rules have been proposed. The first rule is that of a minimum sample of 100 participants as per Boomsma [BOO 82, BOO 85], 150 as per Anderson and Gerbing [AND 88] or Jaccard and Wan [JAC 96], and 200 as per Kline [KLI 16] for a standard model (a not very complex model here). Next is the rule that links the number of participants to the number of free parameters (to be estimated) in the model. Bentler [BEN 95], for example, recommends five times more participants than free parameters when applying the maximum likelihood estimation method or the generalized least squares method, and ten times more participants than free parameters when opting for the asymptotic Distribution-Free estimation method (ADF) or its equivalents. For Jackson [JAC 03], a 20:1 ratio (20 participants for 1 free parameter) will be ideal, while a 10:1 ratio would be acceptable. This rule takes into account both the complexity of the model as well as the requirements of the estimation method. Indeed, it is not uncommon to see the asymptotic distribution free estimation fail when applied to a sample with few participants. A non-positive definite covariance matrix could be the cause after having been the consequence of the small sample size.

It is clear that in both rules the characteristics of the model (complex/simple) are a decisive factor in determining the minimum required sample size. There are others that are just as important: the reliability of indicators of latent variables, the nature of the data (categorical, continuous), their distribution, and especially the type of estimator used, which we have just mentioned (e.g. ML, MLR, WLSMV).

Sample size and statistical power are intertwined such that the former determines the latter, and so the latter is used to determine the former [KIM 05]. Here, let us recall that statistical power refers to the probability of rejecting the null hypothesis (H_0) when it is false. In SEM, as we have seen, the null hypothesis is represented by the model specified by the researcher (i.e. H_0: $F_{min} = 0$, i.e. the specified model fits the data perfectly). Putting this null hypothetical to test, it is important to know the probability of having a good conclusion concerning it (i.e. the probability of accepting a real H_0, and the probability of rejecting a false H_0). The recommended acceptability threshold is a power ≥ 0.80, that is the type II error risk should not go above 20% (or 1 – 0.80). In other words, an 80% probability to not commit a type II error. This happens when the null hypothesis (here, the fit of our model to the data) is accepted by mistake. We know that sample size and statistical power are two important levers allowing for reducing this error.

Several strategies have been proposed to solve the question of sample size and statistical power required for a given structural equation model. For example, Muthén and Muthén [MUT 02b] proposed the use of the rather complicated Monte Carlo method. MacCallum, Browne and Sugawara [MAC 96] introduced another, more practical type of analysis of statistical power and sample size for structural equations models, based on both the fit index, the Root Mean Square Error of Approximation (RMSEA) that we will discuss later, and the number of degrees of freedom of the specified theoretical model (H_0). MacCallum and his collaborators [MAC 96] showed the existence of a link between the number of degrees of freedom and the minimum sample size to reach the acceptable statistical power (0.80). For example, a model showing only 8 *df* (a not very parsimonious model), needs at least 954 participants ($N_{min} = 954$) are needed, while only 435 participants are needed for a model showing 20 *df* (so, a more parsimonious model).

The R "semTools" package has a function (*findRMSEAsamplesize*) using the procedure suggested by MacCallum, Browne, and Cai [MAC 06] making it possible to determine the minimum sample size for a statistical specified *a priori*, based on a hypothetical RMSEA value (for example, 0.05). To determine the minimum sample size for a model, the reader can also use the calculator offered by Daniel Soper at the following address: https://www.danielsoper.com/statcalc/references. aspx?id=89. Based on the approach suggested by Westland [WES 10], this calculator allows you to determine the sample size by taking into account the number of observed and latent variables in the model, and, *a priori*, the effect size, the level of probability (typically $\alpha \leq 0.05$), and the desired statistical power (usually ≥ 0.80). The calculator gives both the required minimum sample size to detect the specified effect and the minimum sample size, taking the complexity of the model into account, in the results.

Let us conclude here that sample size in SEM is a subject that specialists have not finished debating [WES 10, WOL 13], therefore making it impossible to reach a consensus on this subject.

1.3. Model evaluation of the solution of the estimated model

This step corresponds to examine and read the results of the specified model estimation. The confirmatory nature of the approach obviously requires the use of goodness-of-fit indices to judge the compatibility between the specified theoretical model and the data. The only statistical test is χ^2 the use of which is, as we have seen, limited by its sensitivity to sample size. Moreover, structural equation modeling is covered throughout by a paradox: as much as it needs a large sample size, its only statistical test is just that sensitive to it; hence adequacy fit indices that are not statistical tests are used. We can distinguish two types: overall goodness-of-fit indices, which can reject or accept the model applied to the data, and local tests, which make it possible to analytically review the solution obtained (i.e. individual parameter estimates), ensure that it is a proper solution, and determine the significance of the parameter estimates. Let us see them one by one.

1.3.1. *Overall goodness-of-fit indices*

Currently, two dozen are available in all SEM software, including lavaan. But this wealth of indices constitutes the difficulty in selecting the most reliable and appropriate ones, especially as the growing number of books dedicated to them are unanimous neither in their acceptability threshold nor in their meaning and how to interpret them [BAR 07, WIL 11]. Several classifications have been proposed to sort this out, the most commonly known ones being that of Marsh and his colleagues [MAR 88], of Tanaka [TAN 93], and especially that of Hu and Bentler [HU 99], which is still authoritative. Thus, and for the sake of simplification, we chose to group them into three different categories: absolute fit indices, incremental fit indices, and parsimonious fit indices. For each category, we will present the most commonly used indices from the recommendations of Hu and Bentler [HU 99], indicating their level of acceptability and the interpretation given to them.

1.3.1.1. *Absolute fit indices*

These indices are based on a comparison between the observed variance-covariance matrix (S) and the variance-covariance matrix reproduced based on the theoretical model (Σ). The deviations between homologous elements of these two matrices give rise to, as we saw earlier in this chapter, a residual matrix (S – Σ) used to calculate the Root Mean Square Residual (RMR) or the Standardized Root Mean Square Residual (SRMR). Easier to interpret, standardization makes it possible to eliminate the effects of the scale of the variable on the residuals of the model. The more the deviations between the elements of S and Σ are reduced, the lesser the value of the RMR or SRMR. This value is obtained by dividing the sum of each element of the residual correlation matrix squared by the number of variances and covariances of the matrix. The value ranges from 0.00 to 1.00. The fit is better when this value is close to 0.00; but a value equal or less than 0.08 indicates that the model fits the data well. Finally, we will add the Weighted Root Mean Square Residual (WRMR), which is an index suitable for categorical data estimated using the WLS or DWLS method. A value less than 1 indicates that the model fits the data well.

1.3.1.2. *Parsimonious fit indices*

These indices show the originality in taking the parsimony of the theoretical model into account. Parsimony is a central concept in structural equation modeling [PRE 06]. It refers to the small number of parameters to be estimated required to achieve a given goodness of fit. Moreover, it is important to consider that a good model fit as indicated by the fit indices is usually due to either the plausibility of the theoretical representation specified by the researcher or the over parameterization of the model, that is its lack of parsimony. In fact, although it perfectly fits the data, a saturated model has no use because its adequacy is due to its lack of parsimony. It is a complex model with a number of parameters to be estimated equal to the number of variances and covariances observed, and, thus, zero degrees of freedom. In contrast, seen in Figure 1.8, the null model (not to be confused with the null hypothesis), as it is highly parsimonious, is severely disadvantaged. This is a simple model that has little, if any, parameters to be estimated, and therefore a very high number of degrees of freedom, but which fits the data very poorly. The reason is that the scarcity of parameters to be estimated leaves more opportunities for data to freely differ from reproduced data. It is therefore clear that the less parsimonious models tend to approximate data more easily, hence the need to penalize them, otherwise, as pointed out by Browne and Cudeck [BRO 93], there would be great temptation to include meaningless parameters for the sole purpose of making it seem as though the model is a good fit. A goodness-of fit indice that neglects information concerning parsimony can be misleading.

And since the number of degrees of freedom is linked to parsimony, and as it somewhat represents the number of independent and objective conditions according to which a model can be rejected due to the lack of fit, it constitutes an important detail for estimating the plausibility of the model. The parsimony ratio (PRATIO) captures this information. It is the relationship between the model and the number of degrees of freedom of the theoretical model (df_t) and that of the null model (df_n) which is equal to $k(k-1)/2$ where k is the number of measured variables:

$$\text{PRATIO} = \frac{df_t}{df_n} \quad [1.19]$$

But we can get a different ratio by dividing the number of degrees of freedom of the theoretical model by the maximum possible number of degrees of freedom (df_{max}), which is here equal to $k(k+1)/2$. The values of these ratios vary between 0 and 1. A model is even more parsimonious when this value is close to 1. In fact, a saturated model (i.e. the less parsimonious one) gets a zero PRATIO. Some authors, such as Mulaik, James, Van Alstine, Bennett *et al.* [MUL 89), advocate multiplying this ratio with the fit indices to get the adjusted indices taking parsimony into account.

However, the most recommended index in this category is the Root Mean Square Error of Approximation (RMSEA)[9] with its confidence interval of 90%:

$$\text{RMSEA} = \frac{\sqrt{\chi^2 - df}}{\sqrt{df(N-1)}} \quad [1.20]$$

where:

– χ^2 = the chi² value of the specified and estimated theoretical model;

– df = degrees of freedom of the specified and estimated theoretical model;

– N = sample size.

The value of this index ranges from 0.00 to 1.00, with a ≤ 0.06 value indicating that the model fits the data well. Ideally, this value should have a confidence interval of 90%, with a minimum close to 0.00 and a maximum not exceeding 0.100.

In the presence of alternative non-nested models (in competition), some parsimonious fit indices help us discern which of them is the best fitting model. The

9 RMSEA can be classified as an absolute fit index.

Akaike Information Criterion (AIC) as well as the Bayesian Information Criterion (BIC) also perform this function.

The first is obtained in lavaan in the following way[10]:

$$\text{AIC} = -\,2\log L + 2\text{p} \qquad [1.21]$$

where:

– $\log L$ = *log-likelihood* of the model estimated (H_0);

– p = number of parameters to be estimated in the model.

lavaan provides the value of log-likelihood (“logl”) as well as the number of free parameters to be estimated (“npar”).

The second is obtained in lavaan as follows:

$$\text{BIC} = -\,2\log L + 2\text{p}*\log(\text{N}) \qquad [1.22]$$

where:

– $\log L$ = *log-likelihood* of the model estimated (H_0);

– p = number of free parameters to be estimated in the model;

– log(N) = logarithm of the sample size.

When one has to choose between several alternative non-nested models (competitive models), it should be one for which the AIC and BIC values are the lowest, or even negative. But as there are no standard indices, it is impossible to indicate the smallest desirable value as well as the broader unacceptable value [HAY 96].

We clearly see through their calculations whether, for these two indices, the lack of parsimony is to be penalized. Thus, models with a large number of free parameters to be estimated, that is the least parsimonious models (those tending towards model saturation, because over-parameterized), are penalized by an inflation of their AIC and BIC value. The BIC is, from this point of view, more uncompromising than the AIC, and more severely penalizing of the lack of parsimony.

---

10 There are several formulas of this index. For example, in Amos, AIC = χ^2 + 2p. BIC and AIC are calculated by lavaan only with the maximum likelihood estimation method (ML and its “robust” versions).

1.3.1.3. *Incremental fit indices*

These indices evaluate the goodness of fit of the specified model against a more restrictive model considered to be nested in the specified model. The concept of nested models (nestedness) is important and needs to be explained. Two models are considered as being nested when one is a special case of the other. And several models can form a nested sequence when, hierarchically, each model includes the previous models as special cases. In fact, they are alternative models that, while they have the same specifications, differ only in the restrictions that each of them can be subjected to. Such a sequence can be represented on a continuum whose extremes are, on the one hand, the simplest model, that is the null model (M_n) that, undergoing the maximum restrictions, contains no free parameters to be estimated; and, on the other hand, the most complex model, that is the saturated model (M_s) in which the number of free parameters is equal to the number of variances and covariances of the observed variables. The first provides the worst approximation of the data, while the second provides the perfect fit. Figure 1.8 illustrates the same. It will be noted that M_n is nested in M_a; M_n and M_a in M_t; M_n, M_a, and M_t in M_b; and finally M_n, M_a, M_t, and M_b in M_s.

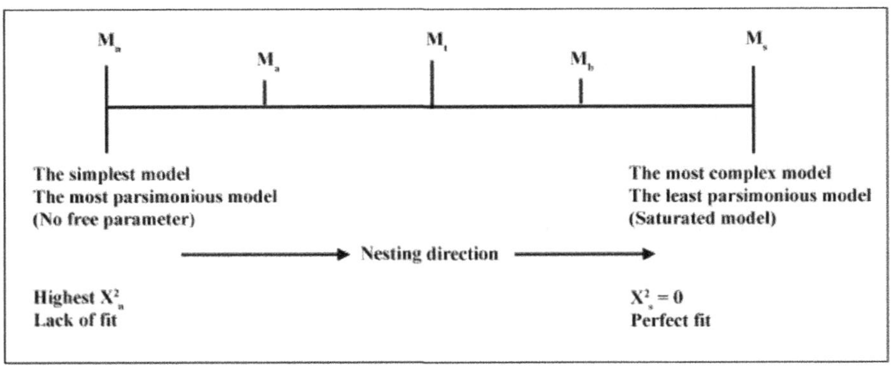

Figure 1.8. *Nested SEM models (inspired by [MUE 96])*

Incremental fit indices are based on the comparison between the specified theoretical model (M_t) and a nested model called the "baseline model", in which the null model (M_n) – also called the "independence model", as it postulates that all covariances are zero, which implies a complete (but unlikely) independence between variables of the model – is the simplest and most widely used version for this type of comparison.

In fact, these indices measure the relative decline of the discrepancy function (F_{min}) (or the relative increase of the fit, which amounts to the same) when we replace a null model (M_n) with a more complex model (for example, M_a, M_t or M_b). Their values range from 0, indicating a lack of fit, and 1 a perfect fit (although they may exceed these limits for some indices). But according to Hu and Bentler [HU 99], a ≥ 0.95 value suggests a good fitting model. However, a value greater than 0.90 is still used for judging whether a model is acceptable or not. A 0.95 value indicates the increase in the goodness of fit of the specified model with respect to the null model: it can indeed be inferred that, applied to the same data, the first model is 95% better than the second. The Comparative Fit Index (CFI) and the Tucker-Lewis Index (TLI) are the best representatives:

$$\text{CFI} = \frac{1 - \max\left[(\chi^2_t - dl_t), 0\right]}{\max\left[(\chi^2_t - dl_t), (\chi^2_n - dl_n), 0\right]} \qquad [1.23]$$

where:

– χ^2_t = the chi² value of the specified and estimated theoretical model;

– df_t = the degrees of freedom of the specified and estimated theoretical model;

– χ^2_n = the chi² value of the baseline model ("null" model);

– df_n = degrees of freedom of the baseline model ("null" model);

– max = indicates the use of the highest value, or even zero if this is the highest value.

$$\text{TLI} = \frac{\left(\frac{\chi^2_n}{df_n}\right) - \left(\frac{\chi^2_t}{df_t}\right)}{\left(\frac{\chi^2_n}{df_n}\right) - 1} \qquad [1.24]$$

where:

– χ^2_t = the chi² value of the specified and estimated theoretical model;

– df_t = the degrees of freedom of the specified and estimated theoretical model;

– χ^2_n = the chi² value of the baseline model ("null" model);

– df_n = degrees of freedom of the baseline model ("null" model).

The advantage of TLI with respect to CFI is that it penalizes the model's lack of parsimony. It always results in fewer degrees of freedom (Figure 1.8). We can recall that the less a model is parsimonious (tending towards saturation) the more likely it is to fit the data. Thus, the model fit is attributable more to the lack of parsimony than the theoretical construction that it represents.

It should be emphasized that, unlike the AIC and BIC indices that are used to compare non-nested models, the CFI and the TLI can be used to compare nested models. We will discuss more on that later.

Which one to choose? The answer is far from simple, as both opinions are still divided about fit indices. Hoyle and Panter [HOY 95] recommend using χ^2 (or scaled χ^2), Goodness-of-Fit Index (GFI)[11], at least two incremental fit indices, in this case the TLI and CFI, as well as parsimonious fit indices when necessary. As for Jaccard and Wan [JAC 96], they propose retaining χ^2, GFI, and SRMR in the category of absolute fit indices, RMSEA in the parsimonious fit indices and finally, CFI in the incremental fit indices.

These opinions are sufficient to show that it is not possible to recommend a standard list of fit indices that could be unanimously accepted, especially since the choice of a fit indice could be guided by theoretical and methodological considerations (e.g. sample size, estimation method, model complexity, data type). In Hu and Bentler [HU 98, HU 99], readers will find details about the sensitivity of these fit indices to all these methodological aspects. The only advice we can afford to give, however, is to closely follow developments concerning fit indices. In the meantime, you can carefully follow the recommendations of Hu and Bentler [HU 99] or Schreiber and his co-authors [SCH 06] who suggest the following guidelines for judging a model goodness-of-fit (based on the hypothesis where the maximum likelihood method is the estimation method): 1) RMSEA value ≤ 0.06, with confidence interval at 90% values should be between 0 and 1.00; 2) SRMR value ≤ 0.08; and 3) CFI and TLI values ≥ 0.95.

However, we will subscribe to the idea of Chen *et al.* [CHE 08] that not only is there no "golden rule" on these fit indices, but there can be no universal, interchangeably ready-to-use threshold for them in all models. In fact, [CHE 08] showed how the universal threshold of 0.05 for RMSEA penalizes (rejects) good models estimated with a small sample size ($N < 100$). And these authors conclude quite rightly about the use of fit indices that *in fine*, a researcher must combine these statistical tests with human judgment when he takes a decision about the goodness-of-fit of the model (p. 491). Table 1.11 summarizes all the goodness-of-fit indices presented in this chapter.

---

11 Proposed by Jöreskog and Sörbom (1974) as early as the first commercial version of LISREL, this first adequacy index (much like the Adjusted Goodness-of-Fit Index, AGFI) was famous for quite long. Highly criticized, it has no longer used.

Fit type	Index	Interpretation for guidance
Absolute	RMR/SRMR	≤ 0.08 = good fit
	WRMR	≤ 1.00 = good fit
Parsimonious	PRATIO	Between 0.00 (saturated model) and 1.00 (parsimonious model)
	RMSEA	≤ 0.05 = very good fit ≤ 0.06 and ≤ 0.08 = good fit
	AIC	Comparative index: the lower value of this index, the better the fit
	BIC	Comparative index: the lower value of this index, the better the fit
Incremental	CFI	≥ 0.90 and ≤ 0.94 = good fit ≥ 0.95 = very good fit
	TLI	≥ 0.90 and ≤ 0.94 = good fit ≥ 0.95 = very good fit

Table 1.11. *Some goodness-of-fit indices available in lavaan*

1.3.2. *Local fit indices (parameter estimates)*

These indices concern all the individual parameter estimates of the model and allow for a more analytical examination of the solution. They make it possible to ensure that it is a proper solution, that is to ensure that it contains no inadmissible values devoid of all statistical sense like standardized estimates higher than 1 or negative variances. These offending values are known as Heywood cases, which have multiple causes. In Dillon, Kumar and Mulani [DIL 87], there are clear details on the sources as well as the solutions for these Heywood cases. Furthermore, the reader may refer to Hayduk [HAY 96] with regard to negative R^2, which are often the ones to suffer in non-recursive models (see Chapter 3 of this book). The author proposes a very interesting solution.

Convergence to a proper solution is compulsory for it to be accepted. Another factor is the coherence of the parameter estimate values from preliminary theoretical considerations. Indeed, to obtain values devoid of any theoretical meaning (as opposed to those predicted, for example) makes a solution nonsensical, even if the overall goodness-of-fit indices point in its favor.

Thus, individually reviewing each parameter is necessary to interpret results, and is useful for improving a model when it fails to fit the data. In fact, it is important to know whether the value of a parameter is statistically significant, to know the error variance (i.e. the unexplained variance portion of a variable) as well as the total proportion of variance of an endogenous variable explained by other variables in the model (R^2). In addition, standard errors, which refer to the precision with which parameters were estimated, must be neither too small (close to zero) nor excessively large (even though there are no precise criteria on this subject).

The significance of a parameter estimate is worth consideration. It refers to using a statistical test to determine if the value of a parameter significantly differs from zero. To do this, we simply divide the non-standard value of the parameter by its standard error ("std.err" in the lavaan results). The ratio obtained is interpreted, when the distribution is normal and the sample large enough, as a *z*-test ("z-value" in lavaan), meaning an absolute value greater than 1.96 helps conclude that the parameter estimate is different from zero at a threshold of statistical significance of 0.05. The threshold of significance of 0.01 requires an absolute value greater than 2.58. However, when the sample size is small with normal distribution, the ratio is read as a *t*-test requiring the use of a table to assess its statistical significance at a fixed threshold. It is useful to know that in the case of a large sample, some very low estimates are often significant.

As we already know, the path coefficient measures the direct effect of a variable that is a predictor of a criterion variable. We also know that there is an effect only when the value of the coefficient is significantly different from zero. What remains to be seen now is the relative importance of an effect.

It is clear that, although statistically significant, standardized path coefficients may differ in the magnitude of their respective effects. Some authors, especially Kline [KLI 16], suggested taking into account significant standardized values close to 0.10 with a weak effect, values that are approximately 0.30 with a medium effect, and those greater than 0.50 with a large effect.

1.3.3. *Modification indices*

Modification indices help identify areas of potential weakness in the model. Their usefulness lies in their ability to suggest some changes to improve the goodness-of-fit of the model. In fact, modification indices (provided by all software) can detect the parameters, which contribute significantly to the model's fit when added to it. They even indicate how the discrepancy function (and therefore χ^2) would decrease if a specific additional parameter, or even a set of parameters (in this case, we use multivariate Lagrange multiplier test; see Bentler [BEN 95] is included. These indices are generally accompanied by a statistic called the *Parameter change*, indicating the value and its tendency (positive *versus* negative) that a parameter would get if it were freed. According to Jöreskog and Sörbom [JÖR 93], such information is useful to be able to reject models, in which the tendency of the values of freed parameters turns out to be incompatible with theoretical orientations. The Wald test, which is to our knowledge available only in the

EQS [BEN 85, BEN 95] and SAS [SAS 11] software, reveals parameters that do not contribute to the model fit of the model and removing which will improve this fit. More detailed and illustrated information on using the Wald and Lagrange multiplier tests can be found in the book by Byrne [BYR 06].

Let us remember that such operations are used with the risk of severely betraying the initial theoretical representation and compromising the confirmatory approach. Whatever it may be, any revision is part of a re-specification approach of the model. We thus move from the actual model evaluation to model specification, and rather model respecification. And a re-specification is all the more sufficient when it is supported by and coherent with the initial theoretical specification. This return to the starting point closes a loop, the outcome of which – after a new model evaluation – is the final rejection of either the model or its acceptance and, consequently, its interpretation.

To conclude, it should be stressed that any solution must be examined in the light of three important aspects: (1) the quality of the solution and the location of areas of potential weakness and misfit (i.e. absence of offending estimates values such as negative variances) and potential areas of weakness of the model (to be identified using modification indices); (2) the overall model fit of the model; (3) local fit indices of the solution (individual parameter estimates and their significance).

Thus, reducing the assessment of a model to its overall fit is a mistake that should not be committed. We will discuss more on that later.

1.4. Confirmatory approach in SEM

In most cases, using multiple regression involves integrating a set of predictor variables in the hope that some will turn out to be significant. And we would have as much chance as the variables introduced in the model. Although guided by some knowledge of the phenomenon in under study, this approach is considered as exploratory. Everyone knows that we often proceed by trial and error, trying various successive combinations, and sometimes getting a little fortunate through this. Moreover, it is this exploratory side that makes this a pretty exciting technique. It is fully covered by the motivation to discover. It is to researchers what excavations are to archaeologists. It is the same with most factor analysis methods whose exploratory nature represents the main purpose.

Structural analyses are part of a confirmatory approach: a model is first specified and then put to the test. A rather simplified representation of reality, a model designates a hypothetical system in which several phenomena maintain various relations with each other. The hypothesis, for which the model is a simple formalization, concerns the exact organization of these phenomena within an explanatory system. It should also be mentioned that apart from the strictly confirmatory approach that seeks to accept or reject a model that is put to the test, we can recognize, with Jöreskog and Sörbom [JÖR 93], two other strategies: the competing models and the model generating, as befits structural equation modeling.

In the case of competing models, researchers specify several alternative models for the same data that they would have specified earlier, in order to retain the best fitting model. For this purpose, they have comparative statistical fit indices that will be discussed later in the book.

Model generating occurs when the initial model fails to fit the data or when a researcher proposes a test model whose limits he knows in advance despite the theoretical contribution that led to its development. The researcher proceeds by way of modification-retest as many times as is necessary to obtain a model that fits its data. This is rather a model generating procedure and not a mdel testing in itself. Each model respecification may theory-driven or data-driven. But, as pointed out by [JÖR 93], the aim of this approach is both the statistical fit of the re-specified model as well as its plausibility and its theoretical likelihood. For, while perfect and statistically adequate, a model has value only if it has an acceptable theoretical plausibility.

It turns out that the model generating strategy is increasingly successful with SEM users. There are at least two reasons for this. The first is the flexibility of this approach with respect to the "all or nothing" advocated by the strictly confirmatory approach. The second is the existence of tests that will help the user examine the model to improve its fit. These tests, the best known of which are the modification indices, the Lagrange multiplier and the Wald tests [BEN 95], are important because they are available in almost all SEM software, and because they are easy to use and interpret even if their effectiveness remains to be proven. The modification index, for example, helps detect parameters that would improve the model fit when included into the model.

The use of post-hoc modifications in model generating approach gives rise to a fundamental problem, that of the confirmatory nature of SEM. Moreover, although these modifications are technically easy to use, their usefulness and effectiveness in generating the right model are still to be proved. The various studies and simulations carried out by MacCallum [MAC 86, MAC 88, MAC 92, SUY 88,], whose fierce hostility towards the model generating strategy is known [MAC 96], show the difficulty in finding the good model, some parameters of which have been deliberately altered for research purposes, despite all the indices and procedures available.

Not that this strategy, which is now popular with the research community, should be banned, but it is important to specify the precautions as regards its use. First, any modification must be supported by theory-based rationale, failing which there is a risk of compromising the original confirmatory approach. Then, it is preferable to opt for competing models that have been previously specified on a clear, theoretical basis. Finally, it is imperative to validate the model generated with a new sample. It is a cross-validation procedure [CUD 89, CUD 83], which also applies to the case where we have a single sample. In fact, it is possible to randomly partition a sample, whose size is large enough, into two parallel subsamples, a calibration sample and a validation sample, or even carry out a multigroup analysis with these two subsamples. There is also an index proposed by Browne and Cudeck [BRO 89], called the Cross-Validation Index (CVI), that helps assess the similarity between models from each of these two subsamples. This index is the discrepancy function (*F*) between the reproduced covariance matrix of the calibration sample Σ_c and the covariance matrix of the validation sample S_v [BRO 93]. Cross-validation is much better when the value of this index is close to zero, indicating that the model is identical in the two sub-samples:

$$CVI = F(S_v, \Sigma_c) \qquad [1.25]$$

Ultimately, the researcher should not lose sight of the confirmatory nature of the approach for which he uses structural equation modeling. It is fully an approach for confirmation of an *a priori* specified model within a hypothetico-deductive reasoning. In this case, the hypotheses are based on a well-defined nomologic network.

Although it is sometimes necessary to ease the strictly confirmatory approach, it is perilous to completely deviate from it. Here, we share the appeal by MacCallum [MAC 95] to the editors of scientific journals, inviting them to reject any article using SEM that does not conform to the confirmatory approach or does not take account of precautions and demands that SEM requires. But we also see that, in reality, the inductive and deductive approaches intermingle constantly, and that the borders between the two are less thick than it seems. In fact, it is rare for even an inductive approach to be atheoretical and completely extrinsic to any hypothesis *a priori*, and for its empirical results to rely on pure laws of chance.

1.5. Basic conventions of SEM

The best way to formalize a theoretical model is to diagramm it. It involves making a graph to represent relationships in a system of variables on the basis of prior theoretical information. It is clear that the causal nature of these relationships can not be demonstrated; however, we refer to causal models (Jöreskog & Sörbom, 1996) and causal diagrams when we deal with path analysis. We will also use "effect" or "effect" to describe the relationships that variables have within a causal model.

There is twofold interest in a diagram. As clear and comprehensive as possible, it should be a faithful translation of the theoretical model; it also establishes an isomorphic representation of all structural equations that can replace the command syntax, thus becoming an input diagram for structural equation modeling software (input) for software through a diagrammer.

Although each software has some specificity, all of them share some conventions concerning diagramming models. Thus, with the beginner user in mind, we opted to present these conventions. This presentation seems fundamental to us, as the diagram is now the emblem of structural equation modeling. It is also essential for fully mastering modeling.

Figure 1.9 shows a standard diagram of structural equation modeling. Here, the modifier "standard" is used to designate basic and simple models with respect to more complex ones that have, for example, correlations between measurement error or feedback loops (i.e. non-recursive models that we will discuss in Chapter 3 of this book).

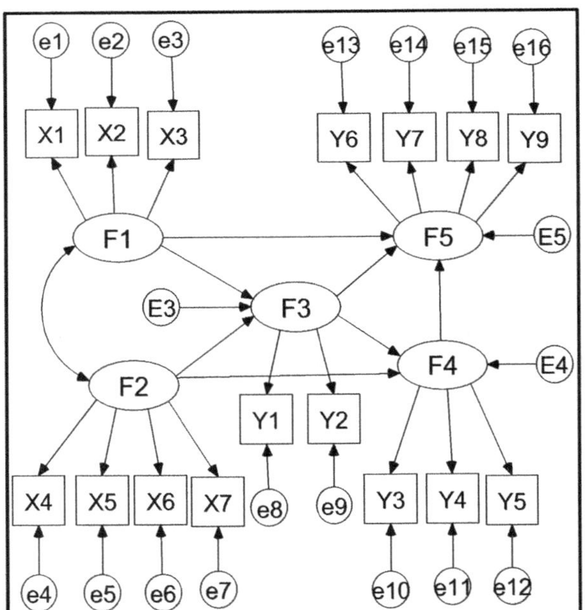

Figure 1.9. *Standard SEM diagram*

It is easily seen that this diagram contains five objects with different facets summarized in Figure 1.10.

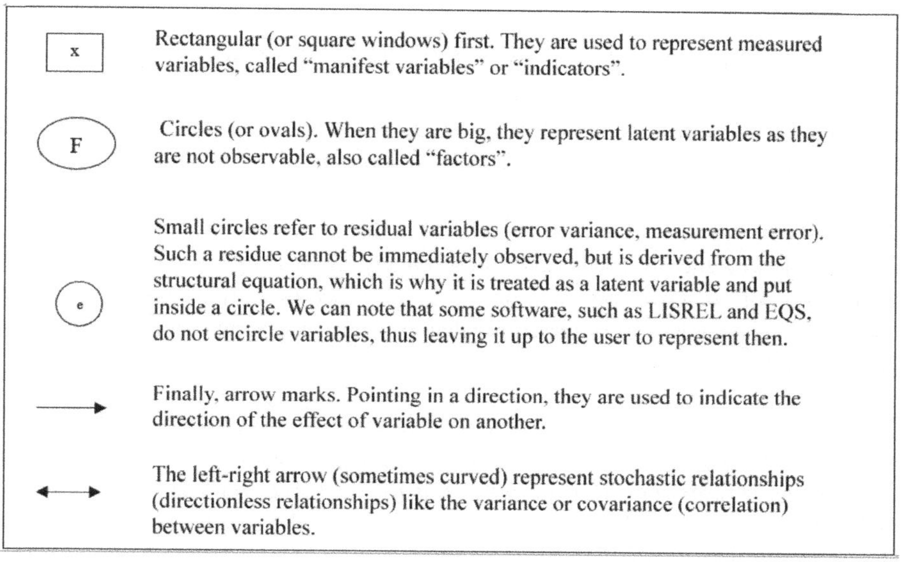

Figure 1.10. *Conventions for drawing diagrams*

1.6. Place and status of variables in a hypothetical model

Whether manifest (observable) or latent, variables can either be exogenous or endogenous. They are considered to be exogenous when they never seem to be criterion variables in any structural equation of a model. They are easily identifiable in a diagram because they have no single arrow pointed at them. Correlated factors F1 and F2 in Figure 1.9 are two exogenous latent variables. Generally, they are the starting point of the model. However, endogenous variables are those that seem like criterion variables in at least one of the structural equations of the model. They are easily identifiable in a diagram, owing to the fact that at least one arrow points to them. Factors F3, F4, and F5 in Figure 1.9 represent endogenous latent variables. An endogenous variable is said to be "ultimate" when it is the last variable in the system. It is in fact the case with F5 in Figure 1.9.

1.7. Conclusion

After having defined structural equation modeling, we tried to show how its foundations are based on regression analysis and factor analysis. By combining them in one approach, structural equation modeling makes them richer and more flexible. And Reuchlin [REU 95] had every reason to write, "the recent structural models are [...] a progress rather than an innovation".

The regression coefficient seemed like a key element in the calculations of structural models. In fact, we have seen its usefulness in the derivation of the matrix that would serve to test the model fit. Model fitting is primarily statistical. It involves knowing whether the dissimilarity between the reproduced correlation matrix and the observed correlation matrix is statistically significant. If such were the case, and with the χ^2 test making it possible for us to confirm it, the model from which the matrix was reproduced is deemed to be unsatisfactory. Otherwise, when χ^2 is not so significant, the question of no statistically significant, the proposed (specified) model provides a good fit to the data. But let us remember that χ^2 is far from being a reliable and robust model fit indice.

Structural equation modeling, which is the generalization of regression analysis and factor analysis, naturally has some points in common with the two. First, all three come under linear statistical models, which are based on the assumption of multivariate normal distribution. The statistical tests used by all these approaches are also based on this assumption. Then, none of them is able to establish a causal link between variables, like any other statistical method. As for the differences between these approaches, three are memorable. The first relates to the approach. It is confirmatory with structural equation modeling and not exploratory, as this is the case with factor analysis or regression analysis. The second lies in the fact that equation models are the only ones able to estimate the relationship between latent variables. The last point concerns the possibility of equation models to insert and invoke mediating variables within a theoretical model. Such an opportunity is not negligible, especially if one accepts like Tolman [TOL 38] that "theory is a set of intervening variables" (p. 344). We will explore these features in more detail in the following chapters.

1.8. Further reading

Readers who wish to deepen their understanding of the main basic concepts in this chapter may refer to the following books:

BOLLEN K.A., *Structural Equations with Latent Variables*, John Wiley & Sons, Oxford, 1989.

FABRIGAR L.R., WEGENER D.T., *Exploratory Factor Analysis*, Oxford University Press, New York, 2012.

GRIMM L.G., YARNOLD R., *Reading and Understanding Multivariate Statistics*, APA, Washington, 1995.

HOWELL D.C., *Fundamental Statistics for the Behavioral Sciences*, CENGAGE, Wadsworth, 2017.

LOEHLIN J.C., *Latent Variable Models. An Introduction to Factor, Path and Structural Analysis*, 3rd edition, Lawrence Erlbaum Associates, New York, 1998.

2

Structural Equation Modeling Software

There is no doubt that the rise of structural equation modeling (SEM) is largely due to the development of many computer programs that are more accessible and less elitist than the pioneer and classic LISREL. In fact, mathematical methods used in SEM are difficult and laborious without the power and speed of computer calculation. The availability of software with more user-friendly interfaces is not specific to SEM. In fact, this trend began only a few years ago and has already created an impact on scientific and office software. But the risk for this lies in the fact that a "ready-to-use" solution would encourage for software mechanical use. Click-button software offer users the impression of having done away with the need to know the theoretical basics of tools and tests that they use. This is not to say that we should discredit these technical advances that have made software more powerful, faster, and more enjoyable to use. However, it should be noted that to address the lack of initiative that these software leave users with, users must seriously invest time on the theoretical foundations of the statistical procedures used. Besides, scientific documentation that accompanies some programs is generally quite complete and, in addition, peppered with examples that are easy to understand and replicate.

Several structural modeling software are currently available. LISREL, the first SEM software [JÖR 74], was succeeded by, among others, EQS [BEN 85], Amos [ARB 95] and M*plus* [MUT 98]. Two innovations have greatly contributed to the flexibility of use of such software: first, the disappearance of the Greek notations that, in LISREL, refer to different matrices and different vectors of a model (Β, Γ, Θ_δ, Λ, Φ, etc.). Second, the command language has been completely simplified (e.g. Simplis for LISREL [JÖR 93]) or completely replaced by diagrams *via* a diagrammer interface (e.g. Amos). Deplored by some, and also hailed by others [BEN 10a], abandoning mathematical symbols in the form of Greek letters, nevertheless, helped to demystify and popularize the SEM. Joeskög and Sorböm [JOE 96], who designed LISREL, justified such a development in order to make modeling accessible to "students and researchers with limited mathematical and statistical statistical training who need to use structral equation models to analyze their data and also for those who have tried but failed to learn LISREL command language" (p. i). What is even more important is the possibility offered by some software in the form of diagrams, and not the command language, to specify the model to be estimated and run the program. The user has to simply enter data, draw up diagrams of the model, and the software takes care of the rest. These programs even provide useful error messages. However, the command language of some software has become so simple that using it has turned out to be more convenient than the GUI, especially in the presence of complex models.

Thus, *grosso modo*, we can distinguish between two classes of SEM software: the commercial and thus paid ones (LISREL, Amos, EQS, Sepath, M*plus*) and those that are free (Mx, OpenMx, sem, lavaan, Onyx, RAMpath) that are open-source. Although still in their early stages (except Mx), they offer features, in constant evolution, comparable to those available in paid software. These free software have the advantage of being all compatible with the R environment. The "sem" package [FOX 06] was the first module dedicated to SEM in R; lavaan [ROS 12] and RAMpath [ZHA 15] are, to our knowledge, the last SEM created in the R environment. Although it has several specificities and certain advances, RAMpath is however backed by lavaan, the presentation and steps for familiarization of which are presented here. Thus, lavaan, an open-source software developing in the R environment, will accompany us throughout this book.

Furthermore, it should be noted that there is a way to use lavaan outside the R environment; this is done by using the JASP environment, an open click-button statistics program (https://jasp-stats.org/). However, the integration of lavaan to JASP, through the SEM (Structural Equation Modeling) module, is in its nascent stage, and hence its use could be temporarily difficult. To open lavaan in the latest version of JASP (version 0.8.4), simply click on the "+" menu of the JASP menu bar and choose "SEM" among the options in the drop-down menu. Let us hope that the JASP environment will soon give lavaan an alternative as complete as the R environment, to which we will devote the following chapter.

2.1. R environment

R is both a mathematical environment and an open-source program or software dedicated to statistically process data [ATT 88]. Designed in 1996 by Ross Ihaka and Robert Gentleman (*alias* "R & R") of the University of Auckland, R is not only free, but also in perpetual evolution as anyone can enrich it as needed by programming new packages and libraries more adapted to the habits, needs, and standards of one's specialty. These include the "ROCR" packages (simplifying the creation of ROC curves for biostatisticians), "R.TeMiS" (intended for sociologists for lexicometric analysis of textual data), and, of course, for this book, "lavaan", dedicated to SEM for a multitude of research specialities

2.1.1. *Installing R software*

The R software is available on GNU GPL (General Public License) on the CRAN website (Comprehensive R Archive Network) https://CRAN.R-project.org. Several mirror sites worldwide provide installation links for the software and packages for additional functions.

2.1.2. *R console*

After the installation is over, we will discover a sober environment made of a tool bar and a simple window with some homepage like information such as the version of the software, guarantee or access to documentation online (see Figure 2.1). This is the R console in which both instructions passed (i.e. commands through which we communicate with the program) and solutions to most of these commands (e.g. the output tables) will be displayed.

If it is possible to enter the command-line directly in the console after the chevron ">", we recommend that the user use the script editor that make it possible not only to build step by step, edit, and comment on the instructions before running them, but also to save all these instructions (or "script") for later use. Finally, the possibility of using copy-paste is another advantage of using the editor, thereby facilitating the implementation of instructions. To open a script, simply click on "File → new script"[1].

1 There are graphical interfaces that are quite user-friendly and ergonomic, such as RKWard (https://rkward.kde.org/Main_Page) and especially RStudio (https://www.rstudio.com/), allowing for easily handling data and making it easier to use R. The interface and usability of the software may be slightly different depending on the software version or the Windows or Mac environment.

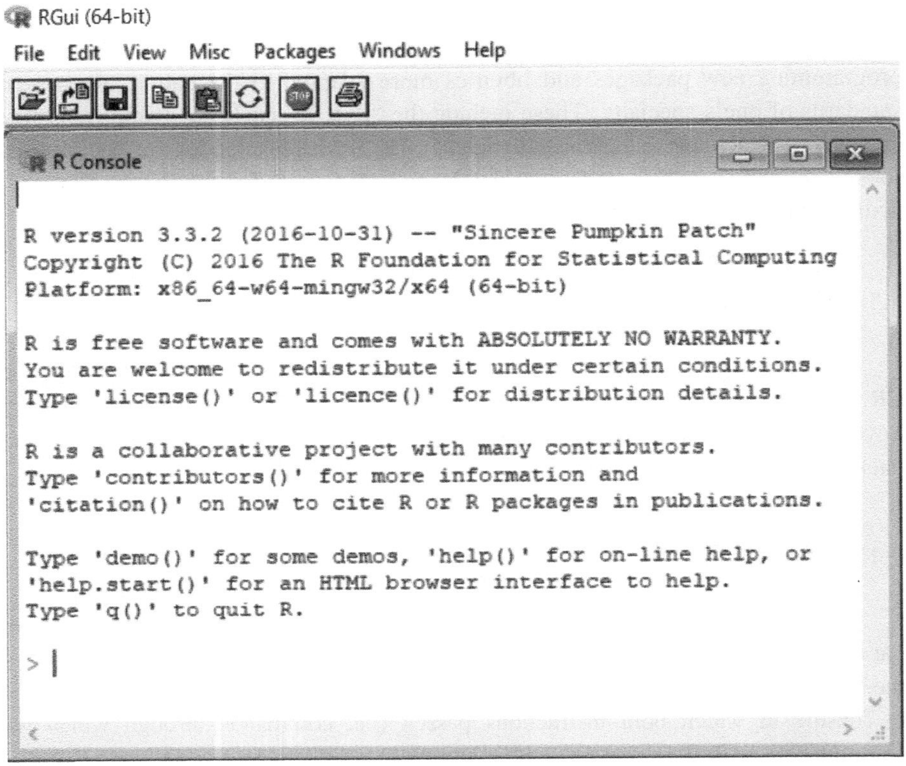

Figure 2.1. *R Console: the startup screen*

Finally, some pop-up windows may appear, such as "Device" containing graphical output and even documentation windows on packages and functions (accessible from "Help" in the toolbar). An overview of the main elements of the R environment is provided in Figure 2.2.

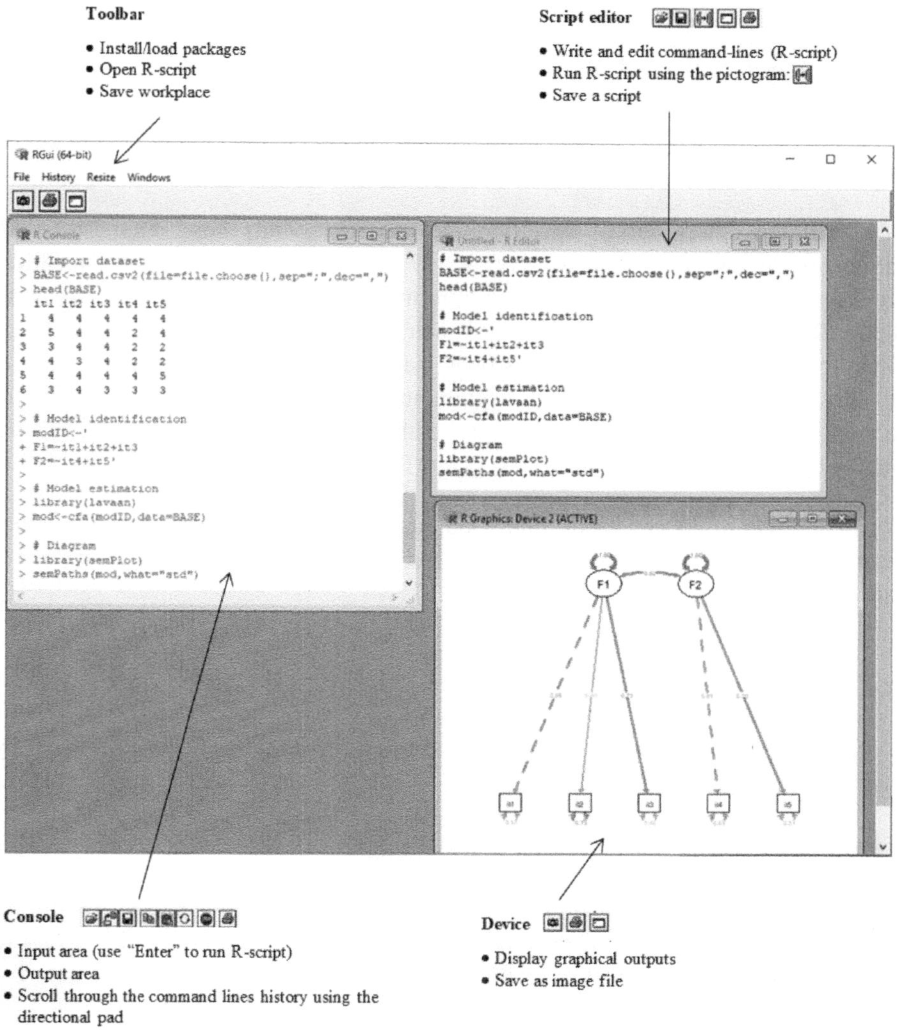

Figure 2.2. *Main components and windows of the R environment*

2.2. lavaan

Developed by Rosseel [ROS 12], the open-source lavaan package has all of the main features of commercial SEM software, despite it being relatively new, as it is still in its beta version, meaning that it is being tested and built. "lavaan.survey" is a very recent add-on.

From the outset, let us specify three essential points here. First, it is not necessary to master R to use lavaan. R will only serve as a mathematical environment for lavaan and, in this case, to import data. However, it is desirable to know R very well to make full use of their mutual and complementary features. Next, communication with lavaan is done using a very simple command language. Finally, lavaan provides an output diagram, meaning a model diagram created from a model analysis. The diagram makes it possible to do a final check of the specified model by the using lavaan syntax (see Figure 2.2).

In addition, users can easily find good quality help and tutorials on the Internet: https://groups.google.com/d/forum/lavaan and lavaan Project (http://lavaan.ugent.be/tutorial/index.html).

2.2.1. *Installing the lavaan package*

Installing a package (for example, lavaan) can be done from the R console by using the following command:

```
>install.packages("lavaan", dependencies = TRUE)
```

It can also be done from the drop-down menu "Packages → Install package(s)…", available in the menu bar of the R console. The user will be asked to select a CRAN mirror site from which the package will be downloaded, only the first time. We can simply select it from the list of packages and double-click to install it (Figure 2.3).

2.2.2. *Launching lavaan*

Once lavaan is installed, it must be launched (opened) each time using the following written command from the R console:

```
>library (lavaan)
```

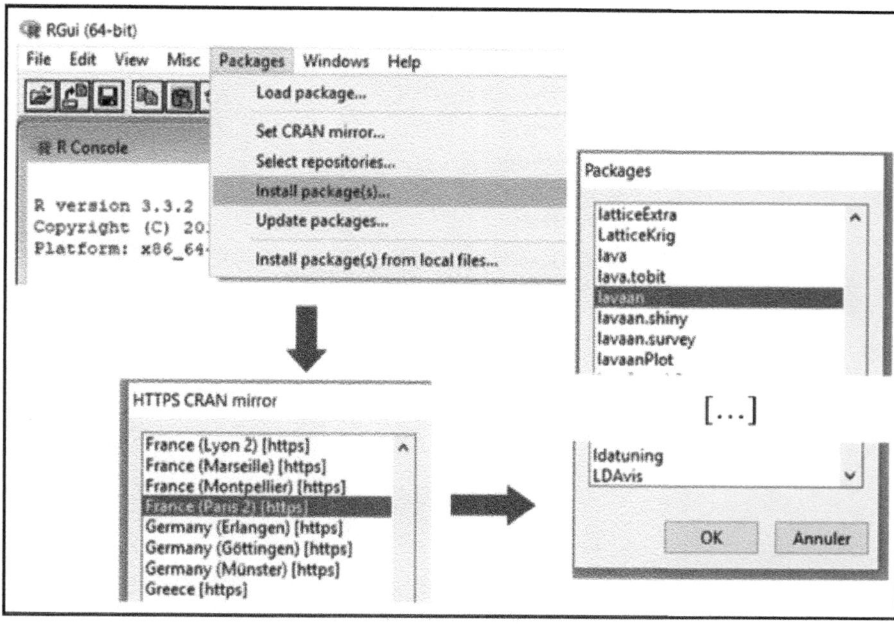

Figure 2.3. *Procedure to install R packages (here lavaan)*

We can also run it by selecting it from the drop-down menu “Packages” and “Load package…”. Figure 2.4 illustrates this procedure. As for lavaan, a message indicates that the package has been successfully activated. Often, nothing indicates that the library is indeed active, except: 1) that there is no error message; 2) R can now access functions logically contained in the package.

It should be noted here that if one wishes to install and/or launch a package through commands written in the R console and not through the menu bar, it is imperative to know that R is: (1) case-sensitive (uppercase and lowercase), (2) sensitive to punctuation marks (comma, decimal point), and (3) sensitive to accentuation (usually to be avoided). On the other hand, spaces have no impact.

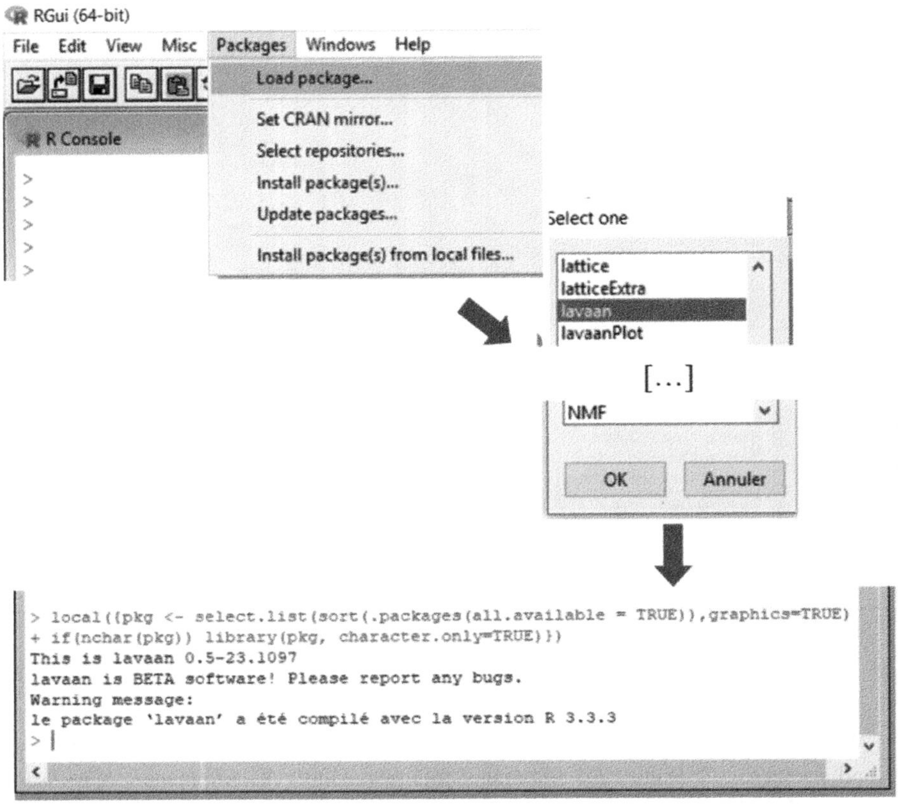

Figure 2.4. *Procedure to load R packages (here lavaan)*

2.3. Preparing and importing a dataset

There are many ways in R to import a data file. We will stick to presenting the method that we consider the simplest, putting ourselves in the shoes of a user who is unfamiliar with the R software. Data is imported in two steps: first, data preparation, and next, the actual import.

2.3.1. *Entry and import of raw data*

The first step is to enter data into an OpenOffice or LibreOffice worksheet, and then save the spreadsheet in the CSV format (separator: semicolon). The file will thus have the extension ".csv". Some precautions have to be taken before saving and importing a file. First, it seems wise to check the presence of outliers quite often due to input errors (for example, entering 44 instead of 4). Second, missing data should preferably be replaced by "NA", meaning "Not Available" (see Table 2.1).

	A	B	C	D	E	F	G
1	it1	it2	it3	it4	it5	it6	it7
2	4	4	4	4	4	4	4
3	5	4	4	2	NA	4	4
4	3	NA	4	2	2	4	4
5	4	3	4	2	2	3	4

Table 2.1. *Spreadsheet saved in text format (.CSV)*

To import the file that we just prepared, simply send the following command from the R console:

```
BASE <- read.csv2 (file = file.choose (), sep = " ;", dec = ",")
```

Apparently complex, this instruction is composed of several simple elements:

– BASE <-: this is the (arbitrary) name given to our data file;

– read.csv2 (): this is the function commanding the reading of a CSV file;

– file = file.choose (): this setting allows for opening a pop-up window offering a way to manually select the appropriate file, which has been prepared in this case (.csv);

– sep = " ;"; indicates that the field separator is a semicolon;

– dec = ",": indicates that the comma is the decimal separator here.

By executing this instruction, a pop-up window opens and allows you to select the CSV dataset to import in R. Figure 2.5 illustrates this procedure.

The base is now imported under the name "BASE". We advise users to carry out at least three checks using "head ()", "names ()", and "str ()" to ensure that the data was properly imported in R (see Figure 2.5).

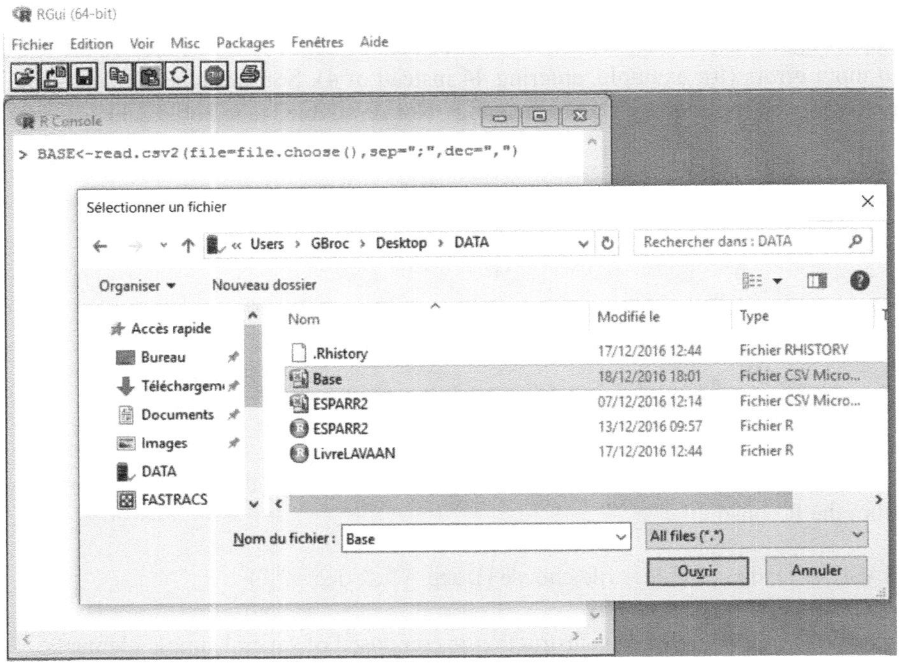

Figure 2.5. *Importing a CSV file into R using "read.csv2" function*

As you can see in Figure 2.6, the "head (BASE)" command is to get the head of the data (BASE), namely the first six observations. It should be noted that the values seem to be properly imported.

The "names (BASE)" command allows for listing the names of the variables available in the imported file (BASE).

Finally, the "str (BASE)" command allows for checking the structure of each variable, which means variables that are supposed to be quantitative are considered by R as such (*int* for "integer" or *num* for "numeric"), and the same goes for variables supposed to be qualitative (*Factor* for "nominal variable" and *Ord.factor* for "ordinal variable").

After the checks are done, the file, arbitrarily called "BASE", is finally ready to be recognized and used by lavaan to test the models that we will have to specify. The following section will discuss how we need to clearly specify these models in lavaan.

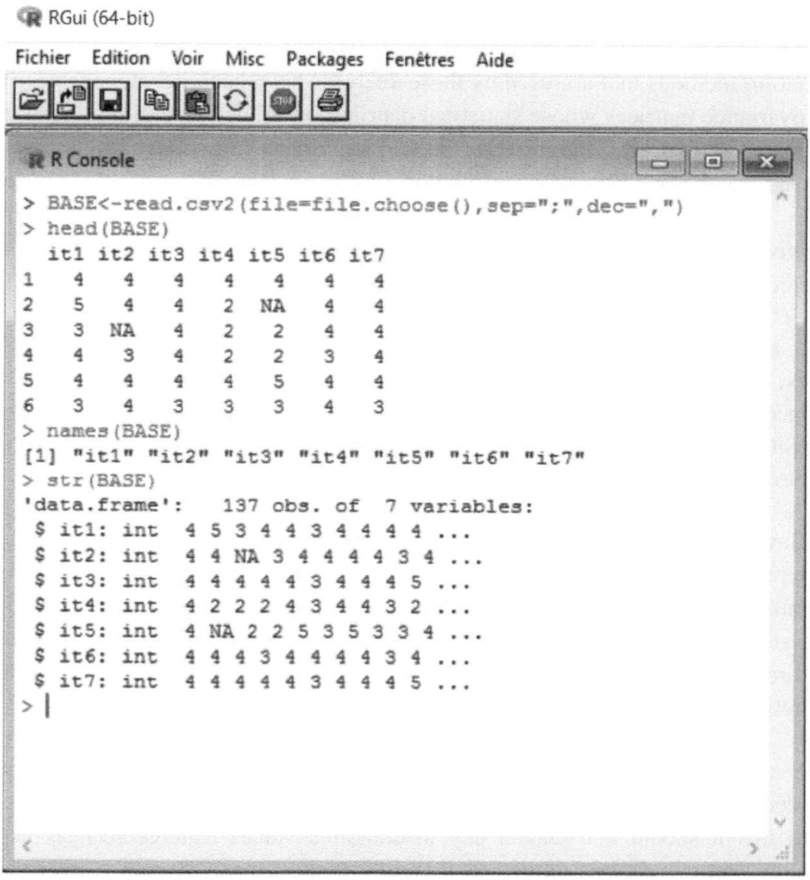

Figure 2.6. *Output of data import verifications*

2.3.2. *What to do in the absence of raw data?*

As we saw, the variance-covariance matrix or correlation matrix constitutes the basic information necessary for SEM. Thus, in the absence of raw data, one of the two matrices is sufficient to estimate some structural models. The question is: which of the two do we prefer?

When we know that the correlation matrix is nothing other than the covariance between standardized variables, it might be surprising that such a question should arise. Let us not forget, however, that the old debate on the interest and the interpretation of standardized and non-standardized coefficients is still not over. Yet, it is usually recommended to use the variance-covariance matrix in structural equation analysis. Incidentally, this analysis often to by the term of analysis of covariance structures? In fact, statistical assumptions underlying conventional estimation methods that are used by these analyses have been developed specifically for covariance matrices whose statistical distribution of elements differs from that of a correlation matrix. Therefore, it seems legitimate to wonder about the risks incurred by the analysis of a correlation matrix instead of a variance-covariance matrix. According to Jöreskog and Sörbom [JÖR 96], such a substitution might not only give incorrect standard errors and affect parameter estimates, but also produces incorrect χ^2 and other goodness-of-fit indices (see also [CUD 89]).

In addition, analyzing the variance-covariance matrix, unlike the correlation matrix, makes it possible to take into account and test hypotheses related to the variances. And finally, the fact that analyzing the variance-covariance matrix makes it possible to get a solution that is both standardized and non-standardized argues in its favor.

However, the interest in analyzing the correlation matrix when the units of measurement of the observed variables are arbitrary is recognized. Similarly, it is recommended that the polychoric correlation matrix (and their asymptotic covariances calculated by most modeling software) in the presence of ordinal or dichotomous measures be analyzed. The reader will find a mathematical presentation and practical illustrations in Muthén [MUT 84, MUT 88, MUT 93].

In any case, it is imperative that the analyzed matrix (but it also concerns the reproduced matrix, the asymptotic covariance matrix) to be positive-definite matrix. This means that it should not contain any inadmissible values features such as negative variances and correlations exceeding the -1.00 to 1.00 range. Surely, a matrix is not recognized as positive definite when the value of its determinant is negative, and as singular when this value is zero. And, since an ill-conditioned matrix cannot be inverted, we then witness the failure of the estimation procedure, especially when it comes to the generalized least squares method or the maximum likelihood estimation method. The causes and remedies depend on the affected matrix, of course. But without going into the details, which are available in Wothkes's seminal contribution [WOT, 93], we can mention a few reasons behind a not positive definite matrix: a high multicollinearity between variables, problems with handling missing data, small sample size etc.

Used as input, these matrices can be accompanied by the observed variable means when the analysis also focuses on means (mean structure analysis). When the analysis focuses on the covariance/correlation matrix (covariance structure analysis), it is assumed that the means of all latent and manifest variables are zero. This hypothesis is not only too restrictive, but also makes us lose a lot of important information. In fact, it is possible and even preferable to combine the two analyses: covariance structure and mean structure. We will come back to this when we discuss multigroup models and latent growth models (see Chapter 4).

Unlike lavaan, some SEM software can analyze a correlation matrix, with the risks noted above, as if it were a covariance matrix. But all software can convert a correlation matrix into a variance-covariance matrix when the standard deviation of the variables are available. The reader will find practical details on how to use a covariance matrix in lavaan or convert correlation matrix into a covariance matrix to be used as input (instead of raw data) in Chapter 3 of this book.

2.4. Major operators of lavaan syntax

Specifying a model is describing the constituting parameters for the software that will be in charge of model fitting procedure (meaning the estimation phase of the model). It involves command-lines containing instructions describing the hypothetical model we want to test. Let us assume that our model is a simple regression of Y on X (Figure 1.2). How do we transcribe this model to lavaan in order for the latter to estimate the former? By using, for example, the tilde[2] (~), which is one of the operators of the lavaan syntax, the following instruction "$Y \sim X$" means that our model is a regression of Y on X. Table 2.2 summarizes the major operators that will be required to transcribe the models into a syntax understandable by lavaan program.

Command	Operator	Illustration	Significance
Estimate a covariance (cor)	~~	X ~~ Y	X is correlated with Y
Estimate a regression	~	$Y \sim X$	Y is regressed on X
Define a reflective latent variable	=~	F =~ item 1 + item 2 + item 3	The F factor is measured by indicators item 1, item 2, and item 3 over which it has effects
Define a formative latent variable	<~	F <~ item 1 + item 2 + item 3	The factor is formed by items 1, 2, and 3

2 To get a tilde: alt gr + 2 (alphanumeric keypad) followed by space or enter (alt + n on a Mac OS).

Estimate the intercept	~ 1	item 1 ~ 1 F ~ 1	Intercept of item 1 Intercept of latent variable F (factor)
Label/fix a parameter	*	F =~ 1*item 1 + b1*item 2 + b2*item 3	Item 1 is set to 1, item 2 is named "b1" and item 3 "b2". The name must begin with a letter.
Constrain parameters to equality	==	b1 == b2	Factor loading of item 1 equals that of item 2 (giving the same name to both items is another way to force them to be equal: b*item 2 + b*item 3).
Create a new parameter	:=	b1b2: = b1*b2	Define a parameter that is not in the model (for example, indirect effect) from the existing parameters. Example: b1b2 = indirect effect of parameters b1 and b2
Insert a comment in the syntax	#	b1b2: = b1*b2 # indirect effect	Explain to the reader the meaning of a command (for example, that here b1b2: = b1*b2 is used to estimate an indirect effect

Table 2.2. *Summary of the main operators of the lavaan syntax*

2.5. Main steps in using lavaan

It should be noted that all the names that precede the assignement operator (i.e. the chevron "<-)"[3] are different names assigned arbitrarily by the user. There are three: a name for the dataset (which may be different from that of the imported CSV file), a name for the specified model, and another for the estimated model. The logic is simple: *in fine*, results concern the estimated model that must be identified by the name assigned to it; the estimation is for the model specified using the syntax and must be recognized by the name it has been assigned, using the data identified by the name assigned to them in step 1. We advise using short, simple names without accentuation. In order to simplify these steps, we will name the imported dataset "BASE", the model to be specified "model.SPE", and the specified model to be

3 The symbol "<-" can be replaced with the "=" symbol.

estimated "model.EST" throughout this book. Table 2.3 summarizes these different steps.

Step	Illustration	Significance
1 - Import data	BASE <- read.csv2 (file = file.choose (), sep = " ;", dec = ",")	Simply import and rename the data file. Here, the assigned name is "BASE".
2 - Model specification	model.SPE <- "model syntax" or model.SPE = "model syntax"	Name the model to be specified. Next, re-transcribe the structural equations contained in a model into a syntax understood by lavaan using operators.
3 - Model estimation	model.EST <- sem (model.SPE, estimator = "ML", data = BASE) or model.EST = sem (model.SPE, ...)	Name the model to be estimated. Proceed to estimating/fitting the specified model named in step 2 by choosing a fitting function (sem, cfa, or growth; see below) using the maximum likelihood estimation method on the data in step 1.
4 - Retrieve the results of the estimated model	summary (model.EST, fit.measures = T, standardized = T, modindices = T)	Obtain the results (standardized coefficients, fit indices, modification indices, etc.) of the estimated model.
5 - View the output diagram of the estimated model	library (semPlot) semPaths (model.EST, "std")	If necessary, obtain the diagram with standardized coefficients. NB. The "semPlot" package should be installed beforehand. There are alternatives like "lavaanPlot" and "RAMpath" packages.
6 - Save the results	saveFile (model.EST, "xxx.txt", "summary")	Save results of the estimated model in a new text file that must be named.

NOTE. – Steps 5 and 6 are optional. As the "lavaan Plot" package is very recent (June 2017), it only gives diagrams of models with measured variables (path analyses) currently. The ability to give diagrams of models with latent variables is soon expected, according to the developers of the package.

Table 2.3. *Main steps for using lavaan*

2.6. lavaan fitting functions

Fitting functions should not be confused with the estimation methods (estimators) referred to above. lavaan has three fitting functions that we will choose at step 3 depending on the nature of the model to be estimated. It is about choosing the analytical framework: confirmatory factor analysis, SEM, or latent growth analysis (Table 2.4).

Fitting function	Illustration	Significance
cfa	model.EST <- cfa (model.SPE, data = BASE)	cfa = confirmatory factor analysis This function is used to carry out a confirmatory factor analysis (estimating a measurement model specified in step 2 and named “model.SPE” here).
sem	model.EST <- sem (model.SPE, data = BASE)	sem = structural equation modeling This function is used to estimate path models and structural equation models with latent variables.
growth	model.EST <- growth (model.SPE, data = BASE)	growth = latent growth modeling This function is used to estimate latent growth models.

Table 2.4. *lavaan fitting functions*

3

Steps in Structural Equation Modeling

An SEM user's toolbox must contain a theoretical model, empirical data and the appropriate software. The mathematical methods used in SEM require such complex and laborious computations that carrying them out manually is a heroic quest! The use of computer programs is more than necessary and they have even proved to be obligatory. Moreover, we have not found any work that uses these methods without using appropriate computation software. The introduction to lavaan in the previous chapter should be enough for one to be convinced that these programs are now quite accessible even to beginners. With regard to the theoretical model, its presence is involved in a slightly different process. Let us recall that this is a confirmatory process that is distinct from the exploratory approach of the data.

This chapter was designed to serve as an introduction to this process that assumes conventions to be observed and sequences to be respected. The objective is to allow the user to go beyond the "technicist" vision of SEM and to better locate its place and importance within a research question. It also proposes familiarizing the user with different statistical indices that are specific to this technique and to help them read and interpret these indices. We will be approaching SEM both from the process point of view (that is, the different steps in using it) as well as the product point of view (that is, its results – more precisely: the solution – and their significance). Figure 3.1 summarizes the different steps in the process within which SEM is used. However, we chiefly focus here on model estimation and model evaluation, that is, examining its solution.

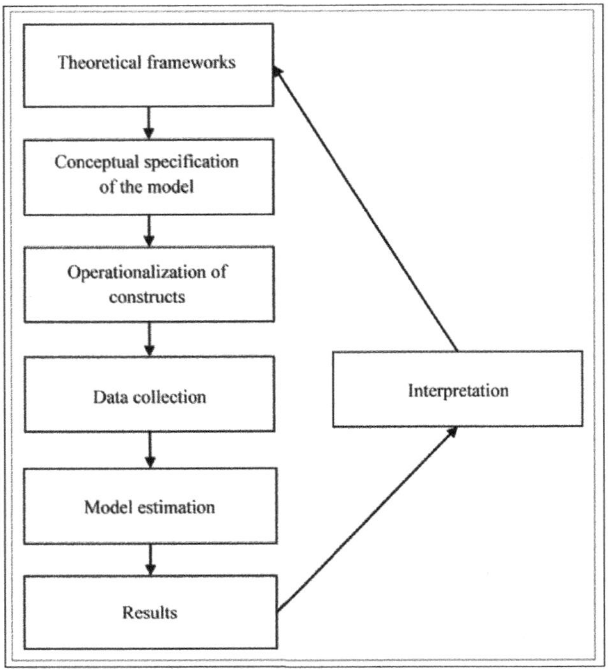

Figure 3.1. *The main steps in application of SEM*

3.1. The theoretical model and its conceptual specification

Being a rather simplified representation of reality, a model serves to formalize a theoretical vision, which is a set of propositions that logically link a number of constructs in order to attempt to explain the phenomena it studies.

Thus, the arrangements of the constructs within a theoretical model results from the hypothetical process and the specification of the research problem. This is much more than a simple proposition relating to the relationships between constructs. "From now on, hypothesis is synthesis," wrote Bachelard in 1934. In the absence of any theoretical *substratum*, we cannot propose a model that discusses predictive links or predictive effects within a set of constructs. The synthetic nature gives the hypothesis its scientific relevance that is its specificity with respect to existing knowledge. This theoretical *substratum* is, here, of a nomological order in the sense that it makes it possible to establish relations between several different constructs (i.e., their roles and status within the model). A perspective of this kind assumes that the measures for constructs used are reliable and valid and that they were developed thanks to robust theoretical definitions of these constructs. However, given the absence here of a theoretical *substratum* for the constructs that we wish to study, how do we select those aspects that we wish to measure among the multitude of other aspects? A measure must always be a precise quantification of a phenomenon that aims to provide a reliable and discriminatory view of it. It is quite obvious today: measurement and definition of a construct form an unique operation and then are the two sides of the same coin.

Once measured, the constructs become variables within a model. In a model, as we have seen, they can be either endogenous or exogenous. Exogenous variables can be recognized by the fact that they are not subject to any predictive effect from any other variable. These are independent variables. The endogenous variables can be recognized by the fact that they are subject to the effect to the effect from at least one other variable, whether exogenous or endogenous. These are dependent variables. An endogenous variable may be considered a mediator variable when it plays the role of an intermediary between two variables. If the mediation (that is, the indirect effect[1]) is effective, then any change in the first variable will affect the mediator variable and any change in the mediator variable will be detectable in the endogenous variable dependent on this. Let us recall that mediation centers on mechanisms that underlie the predictive links between variables [MAC 08]. This endogenous or exogenous characteristic as well as the mediator can also be applied to latent variables, which we will discuss later on in the text. Moreover, as Tolman (1938) wrote, "a theory, as I shall conceive it, is a set of "intervening variables"" that establish relationships between the researchers' theoretical constructs (p. 344).

Irrespective of the theoretical foundations on which it is built, a model is only of interest if it is verifiable, which assumes that we have reliable and valid measures for all constructs that the model will put to the test. By confronting the model with observable data, structural equation modeling makes it possible to evaluate both the likelihood as well as the validity of its measures. Moreover, the confirmation of a model does not really prove its exactitude, only its adequacy, that is how appropriate it is for empirical data. The results from a model are, in part, as valid as the measures used to produce them.

3.2. Model parameters and model identification

A model is a set of parameters that indicate the relationships between a corpus of variables. The parameters are thus the constituent elements of a model.

1 The term "mediation" is inappropriate here as the experimental approach alone is capable of demonstrating the existence of mediation. It would be more prudent to talk about an "indirect effect", which is a necessary, but not sufficient, condition to establish mediation. We will return to this point.

Parameterizing a model consists of distinguishing between its constituent elements in order to communicate them to the software that will take charge of estimating them. In a model the value of a parameter can be free, fixed or constrained. A free parameter is a parameter whose estimation is to be obtained. On the other hand, a parameter is fixed when it is assigned a value *a priori*. A parameter set to zero simply signifies that it is removed from the model. Finally, a constrained parameter is one that is free but obliged to be equal to one or several other parameters. This is an equality constraint between parameters.

The number of parameters no, to be estimated in a model play a role in its identification. In effect, before proceeding to estimate the free and/or constrained parameters, it is necessary to ensure that the model be identified. Model identification is a fundamental operation in SEM. Here, we have chosen to focus more on the logic underlying its use rather than its multiple and complex mathematical aspects. The reader can find a more technical discussion in Bollen [BOL 89] and Rigdon [RIG 95].

In fact, the concept of identification refers to two complementary aspects. First, it refers to the unique nature of a solution: an identified model is one that must yield the unique solution for each of the specified parameters. This sometimes involves ensuring that there is no equivalent models that coud fit the data well. To clarify this, let us take a simple and oft-used example for illustration. Let us assume that a theoretical model suggests a certain value for $X + Y$, and the data suggests that $X + Y = 6$. It is clear that there is no unique solution to resolve this equation. One solution could be a value of 4 for X and 2 for Y ($4 + 2 = 6$); another could be a value of 3 for both unknowns ($3 + 3 = 6$). To obtain a unique estimation for each of the unknowns it is necessary to impose restrictions: for example, let us set the value of X to 1.00, which allows us to obtain a unique value for Y, that is, 5, to resolve the equation.

The identification also refers (and the above equation shows this) to the number of free parameters (i.e., unknowns) with respect to the quantity of data available. Here, the data available to us are generally the variances and covariances of measured variables. The following formula makes it possible to calculate their number: $k (k + 1)/2$, where k is the number of observed variables. The problem that then arises is that of knowing whether we have enough information relative to the measured variables in order to estimate all the unknowns in the model. Three cases may emerge:

– the model is just-identified. We speak of a "just-identified" or "saturated" model when the number of variances-covariances of the observed variables is equal to the number of free parameters in the specified model. With the number of the degrees of freedom being the difference between the first and the second (see [1.14], Chapter 1), a saturated model is a model with zero degrees of freedom. Although this fulfils the minimum condition for identifiability, such a model is only of limited scientific interest as it is never statistically rejectable;

– the model is under-identified. A model is said to be under-identified when the number of variances-covariances of the measured variables is lower than the number of parameters to be estimated in the specified model. Such a model suffers from a deficit of information that is necessary to determine the value of each free parameter. Moreover, the negative number of degrees of freedom that it displays makes this model unacceptable. It is recommended that constraints (for example, setting one or more parameters to zero) should be introduced in order to identify such a model;

– the model is over-identified. A model is said to be over-identified when the number of available variances-covariances is greater than the number of parameters to be estimated. Consequently, the number of degrees of freedom becomes positif, thus restoring to the model its heuristic character. In effect, contrary to the just-identified model, an over-identified model is testable, even running the risk of being rejected. However, as essential as this over-identification is when it brings enough information to estimate the free parameters, it is also prone to affecting the overall evaluation of the model, as we then come up against the concept of parsimony, the details of which will be presented further on in the text.

3.3. Models with observed variables (path models)

Designed at the beginning of the 20th Century by the American geneticist Sewall Wright [WRI 21, WRI 34] to estimate and distinguish between hereditary effects across generations, path analysis was the forerunner of SEM. Along with informatization through means of the computerization of calculations and the confirmatory nature of the process, two advantages over multiple regression contributed to the development of path analysis: its multivariate approach (i.e. the possibility of processing several dependent variables simultaneously) and, most importantly, the possibility of decomposing the total effect of one variable over another into direct and indirect effects. The main characteristic, also being the main limitation, of these models is that they specify relationships between measured (observed) variables only. From such models arise the rather implausible hypothesis that the measures of the variables or constructs are free of measurement errors.

Since Wright and the beginning of SEM, any reference to causality raises hackles even today among some purists, despite the precautions and clarifications brought in and repeatedly stated by Wright [WRI 21, WRI 60] himself, as well as by Duncan [DUN 66].

Remaining faithful to the spirit and the recommendations of these pioneers, we will continue to speak of "causal paths" or "causal effects", which some prefer to call, as a precaution or to satisfy purists (a bad reason), "predictive path models" and "predictive effects". Moreover, we will use them interchangeably here. Yet others prefer putting quotation marks around the designation, "causal paths" to highlight the strict significance (experimental) and to express the caution they wish to demonstrate.

3.3.1. *Identification of a path model*

This diagram is useful to understand the logic behind the identification of a path model and to comprehend its advantages. Let us take the model illustrated in Figure 3.2. It can be seen that this model makes use of five measured variables, three of which are exogenous and correlated with each other (X1, X2 and X3) and two of which are endogenous, predicted (among others) by the first three. We can thus detect six causal paths (P1, P2, P3, P4, P5 and P6) indicating six different direct predictive effects: these are, respectively, the effect of X1on Y2, that of X2 on Y2, of X2 on Y1, of X3 on Y1, of X3 on Y2 and, finally, the influence of Y1 on Y2. We also discover two indirect predictive effects: that of X2 on Y2 *via* Y1, and that of X3 on Y2 *via* Y1. We finally have the residual variables 'el' and 'e2': the first represents the part of the variance in Y1 that cannot be imputed to X1, X2 or X3 and the second represents the part of the variance in Y2 that cannot be imputed to any of the variables. As it is next to impossible to be able to explain the entirety of the variance of a dependent variable, we introduce a residual variable "e", which represents the effect of the variables that are not included in the model. Given that this residual variable is not directly observed but is derived from the equation, it will be placed in a circle and considered as a latent variable. A similar strategy is used for the error term in the regression analysis, the sources of fortuitous variations, that is, the proportion of variance that is not explained by the variables in the model.

In reality, Figure 3.2 is incomplete even if it corresponds perfectly to the typical diagram of SEM that we may encounter in literature. In effect, certain authors require not only that residual variables be encircled, but also the variances ('v' in Figure 3.3) of all the exogenous variables be represented explicitly in the diagram as they consitute effective free parameters in a model. Figure 3.3 presents one such configuration. This presentation offers the advantage of revealing all the free parameters to be estimated. Indeed, each exogenous variable has a variance to be estimated. We can see that there are five of these, with two variances being linked to the residual variables (el and e2). These are, quite evidently, considered as exogenous variables as there is no arrow pointing towards them. Moreover, for practical reasons generally linked to the lack of space, these variances (or even the residues) are often implicitly presented in the diagram, which makes model identification of the model all the more difficult for beginners.

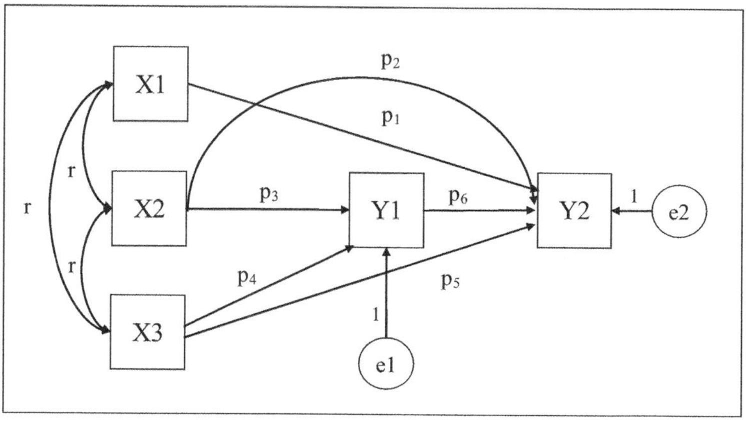

Figure 3.2. *Path model relating five observed variables*

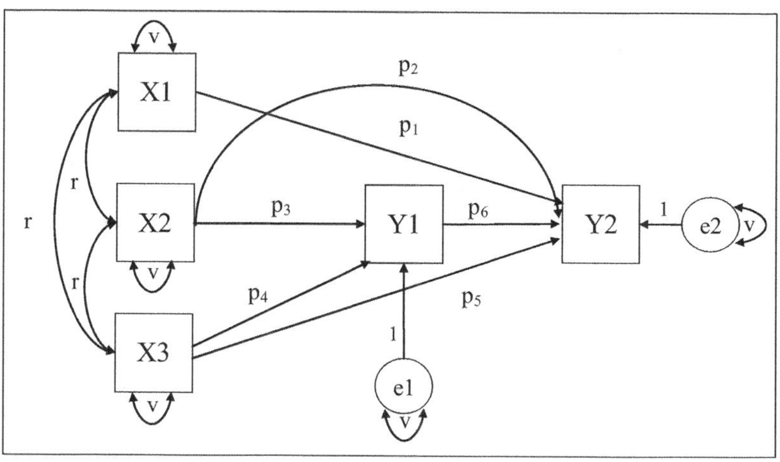

Figure 3.3. *Path model relating five observed variables (the curved arrows with two heads represent variances and covariances/correlation)*

Is the model illustrated by figures 3.2 and 3.3 over-estimated and, therefore, apt to be estimated? To know this, let us calculate the number of its degrees of freedom (*df*) using the following equation (see Chapter 1 in this book):

$$df = [k\,(k + 1)/2] - p \qquad [3.1]$$

where

– (k) = the number of observed variables;

– (p) = the number of free parameters to be estimated.

The model represented in Figure 3.3 has five measured variables and 16 parameters, two of which are set to 1.00, in order to identify the model (this is because we are more interested in the variance of the residual "e" than its effect). The 14 free parameters indicated: 6 causal paths (P1 to P6) + three coviarances between the exogenous variables (denoted by an *r* on the graph) + five variances (denoted by the double-headed curved arrows with a (v)):

$$df = [5(5 + 1)/2] - 14$$

$$= 15 - 14$$

$$= 1$$

Notwithstanding the narrowness of this *df*, which poses parsimony problems in the model and which we will discuss further on, our model is over-identified and, therefore, ready to be estimated.

3.3.2. *Model specification using lavaan (step 2)*[2]

To specify a model is to retranscribe it for the software by using a syntax where operators are symbols. The retranscription concerns structural equations and the parameters that a model counts. Let us recall that there are as many equations as there are endogenous variables in the model; in an equation, there are as many terms as there are arrows directed towards the endogenous variable; and relationship between variables is expressed as linear function whose path coefficient (P) is the value. For example, in the model illustrated by Figure 3.3, there are two endogenous variables (Y1 and Y2). Consequently, there are two equations:

Y1 = X2 + X3 + e1

Y2 = X1 + X2 + X3 + Y1 + e2

2 See Table 2.3.

A look at Figure 3.3 reveals that Y1, for example, has three incoming arrows: one from X2, another from X3 and the third from "e1", this last being set to 1.00 in order at 1 in order to identify the model.

Retranscribing the entire model in Figure 3.3 into lavaan syntax is startlingly simple. This is step 2, which follows data import. As stated, throughout this book, all models to be specified will be named "model.SPE" and all models to be estimated will be named "model.EST". The data imported in step 1 (see the different steps for the use of lavaan, described in Table 2.3 in Chapter 2) will be called "BASE". Let us recall that the tilde (~) is the operator required for the specification of regression models, including path models. We should also recall that the syntax that specifies a model must be placed between straight quotation marks (' ') and not the curved quotation marks usually used in typography (‘ ’). It will be seen that we use the symbol '<-' even though it is possible to replace it by the symbol '='. We finally highlight the fact that the presentation of each step as well as the comments that follow # are not part of the commands but are used to explain them or add observations. Here is an illustration of this:

Step 2. Model specification (Figure 3.3).

```
model.SPE <- 'Y1 ~ X2 + X3
              Y2 ~ X1 + X2 + X3 + Y1'
```

This syntax translates two regression equations. The first equation translates the regression of Y1 on X2 and X3, while the second translates the regression of Y2 on X1, X2, X3 and Y1. We can note the absence, in this syntax, of the residual variables (e1 and e2) fixed automatically and by default at 1.00. Above all, we can note the absence of covariances between the exogenous variables, recognized by default by lavaan. To derogate from this, it is enough to ask for this in model specification.

For example, if we wished to delete the correlation between X1 and X3, the lavaan syntax for this model would be as follows:

STEP 2. Specification of the model in Figure 3.3 without correlation between X1 and X3.

```
model.SPE <- 'Y1 ~ X2 + X3
              Y2 ~ X1 + X2 + X3 + Y1
              X1 ~~ 0* X3' #No correlation between X1 and X3.
```

Finally, it is possible to separate the specified equations (commands) using a semi-colon (;):

STEP 2. Model specification.

```
model.SPE <- 'Y1 ~ X2 + X3; Y2 ~ X1 + X2 + X3 + Y1; X1
~~ 0* X3'
```

3.3.3. *Direct and indirect effects*

Breaking down the total effect into two additive parts, the direct and the indirect effects, made it possible to arrive at a finer assessment of the relationships between variables. We talk about a "direct effect" (DE) when a variable directly predicts another variable. An oriented relationship in a diagram indicates the direction of this prediction, the value of which is its regression coefficient. It is easy, for any endogenous variable, to determine the number of direct effects acting on it. This number simply corresponds to the number of arrows pointing towards it. In other words, in a model there are as many direct effects as there are causal paths. An indirect effect (IE) comes about when a variable predicts another and when this second variable in turn predicts a third. We then assume that the first variable has an indirect predictive effect on the third, through the intermediary of the second variable. This example is the simplest form of mediation; it could, of course, be more complex.

However, we cannot stop here in this decomposition as a close study of Figure 3.3 reveals another source for effects. As this is neither direct nor indirect, it can be imputed to the covariances between the exogenous variables (X1, X2, X3) in the model. Indeed, it is clear that the interdependence between these variables comes into play each time that one of them is engaged in a direct relationship with another variable. This effect can be called the non-predictive effect (NPE) or even a spurious effect. Let us take the example of the relationship between X1 and Y2, which is expressed through the direct effect that the first has on the second, and also through the associations that X1 has with X2 and X3.

The direct effect of one variable on another corresponds to the path coefficient that connects them. As we have seen, this is a regression coefficient, often estimated using the method of least squares or the method of maximum likelihood (which is almost the same thing, here). To return to the example presented in Figure 3.3, the indirect effect of X2 on Y2 corresponds to the product of the two path coefficients (p3 and p6). Thus, the total effect is the sum of all direct and/or indirect effects of the independent variable on the dependent variable.

In a path model, the decomposition of the correlation (or covariance) between two variables makes it possible to distinguish between the direct and the indirect effect of the independent variable (X) on the dependent variable (Y). This decomposition can be carried out using the following formula:

$$r_{YX} = \Sigma_q P_{Yq} r_{qX} \qquad [3.2]$$

where:

– q denotes all the variables in the model that have a direct relationship with the dependent variable Y;

– P_{Yq} refers to the path coefficient that links a variable q to the variable Y.

To apply this formula to the relationship between X2 and Y2, for example, in the model in Figure 3.3, we can proceed as follows.

Identify all variables, q, that have a direct effect on the dependent variable Y2. In this case, we have X1, X2, X3 and Y1.

Identify the path coefficients, P_{Yq}, corresponding to the relationships identified above. In this case, we have: p1, p2, p5 and p6.

Multiply each of these coefficients by the correlation between X2 and each of the variables q. We obtain the following:

$p1r_{X2X1}$

$p2r_{X2X2}$

$p5r_{X2X3}$

$p6r_{X2Y1}$

Upon adding the above products:

$$r_{X2Y2} = p1r_{X2X1} + p2r_{X2X2} + p5r_{X2X3} + p6r_{X2Y} \qquad [3.3]$$

As r_{X2X2} = 1 (it is different for a covariance) and as r_{X2X1} as well as r_{X2X3} are stochastic relationships that cannot be decomposed, a single unknown, r_{X2Y1}, remains to be resolved in [3.5]. In order to do this, we apply [3.4] with the steps that we just described. Thus:

$$r_{X2Y1} = p3r_{X2X2} + p4r_{X2X3} \qquad [3.4]$$

$$= p3 + p4r_{X2X3}$$

Finally, using [3.3] and [3.4] to resolve [3.2], we obtain the following:

$$r_{X2Y2} = p1r_{X2X1} + p2 + p5r_{X2X3} + p6\ (p3 + p4r_{X2X3}) \quad [3.5]$$

$$= p1r_{X2X1} + p2 + p6\ (p3 + p4\ r_{X2X3}) + p5r_{X2X3}$$

$$= p1r_{X2X1} + p2 + (p6*p3) + (p5*r_{X2X3}) + (p6*p4*r_{X2X3})$$

$$= DE + DE + IE + NPE + NPE$$

While this is not complicated, the procedure that we just described does pose the risk, if it is manually computed for a complex model, of forgetting a path or a parameter. It is futile to describe more of this decomposition procedure as it can be carried out by most modeling software. We would, however, refer the reader to work by Bollen [BOL 87] and Sobel [SOB 87] for a more detailed discussion of this aspect of structural equation models.

Finally, it would be useful to enphasize that mediation assumes the existence of a change. It assumes the existence of a causal mechanism that would explain this change; a mechanism through which a variable causes a change in another variable, which, in turn, causes a change in a third variable. If there is No change no mediation. As it is impossible to measure change through cross-sectional study, it would be more cautious to talk about an "indirect effect" than to talk about "mediation". An indirect effect is a necessary, but not sufficient, characteristic of mediation.

3.3.4. *The statistical significance of indirect effects*

Direct effects (i.e., path coefficients) significance tests are available, however, it becomes all the more necessary to use complementary tests to judge the significance of indirect effects [SOB 87]. Several procedures were proposed to evaluate the mediation effect (the indirect effect) of a variable [MAC 08, MAC 02]. There is the "causal" steps procedure proposed by Baron and Kenny [BAR 986], that of the product of coefficients, the distribution of the product of two random variables or again, the bootstrap procedure. Contrary to the other procedures, bootstrapping neither requires a normal, multivariate distribution, nor a large sample. MacKinnon, Lockwood and Williams [MAC 04] estimate that it produced results (confidence intervals) that are more precise than those from other procedures. This procedure is based on a simple principle [MAC 08]. A certain number of samples (fixed by the researcher; for example: "bootstrap = 1000") are randomly generated with replacement starting from the initial sample considered, like the population. Each generated sample contains the same number of observations as the initial sample.

There will then be as many estimations of the model as there are generated samples. The mean of each parameter estimate is computed and accompanied by its confidence interval (CI). An estimation (the indirect effect, for example) is significant at 0.05 if its confidence interval at 95% (95% CI) does not include a null value (see [PRE 08a]).

It is important to note that a serious limitation of lavaan is that it does not automatically compute the indirect effects and the total effect. In order to obtain these effects, a rather convoluted process must be used. We begin by naming the different causal paths (for example, p1, p2, p3, etc.) and we then use the operator ":=" to estimate the indirect effects. Here is an illustration of this for the model 3.3. in which there are four direct effects (labeled p1, p2, p5, p6) and two indirect effects (labeled p3 and p4) on the ultimate endogenous variable Y2:

STEP 2. Model specification.

```
model.SPE <- 'Y1 ~ p3*X2 + p4*X3

# Specify the four direct effects.

  Y2 ~ p1*X1 + p2*X2 + p5*X3 + p6*Y1

# Specify the two indirect effects.

p3p6:= p3*p6
p4p6:= p4*p6

# Specify the total effect

total:= p1 + p2 + p5 + p6 + (p3*p6) + (p4*p6)'
```

The bootstrapping procedure is carried out during the step of model estimation (step 3).

3.3.5. *Model estimation with lavaan (step 3)*

lavaan offers three model-fitting functions: (1) "sem" (for structural equation modeling) to estimate path models and general structrural equation models; (2) "cfa" (for confirmatory factor analysis) to estimate measurement models; (3) "growth" to estimate *latent growth models*.

The parentheses that follow the selected model-fitting function –"sem ( )", "cfa ( )", growth ( )" – must compulsorily contain the name of the model specified in step 2 as well as the file with data named in step 1 "data = xxx". When needed, we can insert other options such as "ordered = c ( )", to indicate, between parentheses, the binary or ordinal variables in the model; "orthogonal = TRUE" to estimate a measurement model (CFA) whose factors are not correlated with each other; "se = "bootstrap"" to obtain the confidence interval for the parameter estimates for the parameters at the end of 1000 resampling procedure (bootstrap = 1,000). The results of the bootstrap operation will be interpreted further in the chapter.

STEP 3. Model estimation using the the model-fitting function "SEM".

```
model.EST <- sem (model.SPE, data = BASE, se =
"bootstrap", bootstrap =1000)
```

STEP 4. Retrieving of the results of the estimated model.

```
summary (model.EST, fit.measures = TRUE,
standardized = TRUE, modindices = TRUE, rsq = TRUE)
```

3.3.6. *Model evaluation*[3] *(step 4)*

The evaluation of the estimated model is carried out by examining the solution or the output, the results yielded by lavaan. The content of these results depends on the elements in the "summary" function brought in step 4, which is relative to the retrieving of the results of the estimated model. For example, the elements within parentheses after "summary" request the results of the estimated model, named "model.EST" by providing goodness-of-fit indices (fit.measures = TRUE), the standardized estimates (standardized = TRUE), the modification indices (modindices = TRUE), as well as R^2 for the endogenous variables (rsq = TRUE).

Let us recall here that the evaluation of a solution must be carried out with respect to both three aspects: 1) the overall model fit; 2) the quality of the solution and the location of areas of potential weakness and misfit; and 3) local fit indices of the solution (i.e., parameter estimates).

---

3 In order to do away with exponential notations in the results, it is enough to introduce the following command: "> options (scipen = 999)" and to then launch the "summary" instruction described earlier. For example, p = 8,083672e – 23 becomes: p = 0,000000000000 0000000008083672.

3.3.7. *Recursive and non-recursive models*

We generally distinguish between two types of structural equation models (including path models: recursive and non-recursive models. A recursive model is one in which no endogenous variable has any effect on itself. The second, on the other hand, is a model in which an endogenous variable may have an indirect effect on itself. This is a model in which a reciprocal effect may be produced between different variables: the first influences the second, which in return influences the first[4]. We speak here of causal reciprocity or reciprocal predictions. This is illustrated by Figure 3.4.

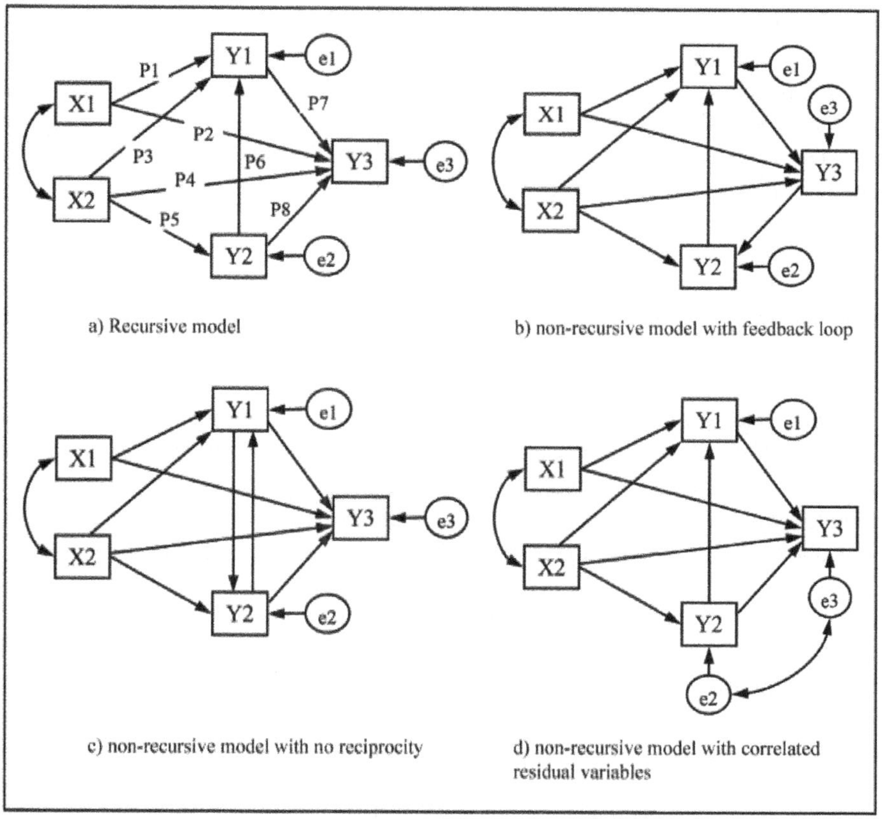

Figure 3.4. *A recursive model and three types of non-recursive effects*

4 The lecture may be rightly surprised that these models are incorrectly called "non-recursive" when the adjective "recursive" would appear more appropriate (from the Latin *recursum*, de *recurrere*, "to go backwards").

We can easily see that the only difference between diagrams (a) and (c) in Figure 3.4 is at the level of the relationship between the variables Y1 and Y2; this is a unidirectional relationship, going from the second to the first, in the diagram (a), and bidirectional (that is, reciprocal) in diagram (c). We can also observe that unlike a covariance relationship – between X1 and X2, for instance – this reciprocity is reflected by two distinct paths, one expressing the influence of Y1 on Y1 and the other expressing the reverse influence. This model could be particularly useful in determining the direction of prediction and evaluating opposing approaches, such as the *bottom-up versus top-down*, that we encounter in various fields of study [WON 99]. Nonetheless, from a conceptual point of view, the causal reciprocity is debatable as it hypothesizes a complete absence of temporal precedence between the cause and the effect. The simultaneity of the effects of two variables on one another does away with the principle of existence of a certain amount of time that is necessary for the production of any causal effect. Any causal relationship requires an interval of time, however infinitesimal, for producing the causal effect. Thus, can one variable be the cause of a second which itself is the cause for the first? Has the causal effect already come into play in order to produce an immediate retroactive effect? It is reasonable to question this.

Causal reciprocity is not the only type of non-recursive effect; the feedback loop, illustrated in Figure 3.4b is another type. In this model, a loop is formed between Y2, Y1 and Y3: the first variable influences the second, which influences the third, which, in turn, influences the first. Another type of non-recursive effect, illustrated in Figure 3.4d is expressed in a model through the correlation between residual variables. However, not all researchers agree on the non-recursiveness of a case like this [RIG 95]. Moreover, there is no non-recursiveness when the endogenous variables with which the residual variables are correlated have no links between themselves. Such a model may, at most, be considered as being partially recursive.

Its limitations notwithstanding, non-recursiveness adds a dimension to path analysis that makes it a remarkable method, posing a serious challenge to multiple regression analysis. The reader may observe and be rightfully surprised that these models are not very frequently mentioned in literature. They will know that these models often encounter identification problems, thereby making it impossible to estimate them. These problems come about even when the number of variances-covariances is greater than the number of free parameters in the model. As this condition is necessary but not sufficient, other conditions are thus required here: order condition and rank condition. In Bekker [BEK 94], Kaplan, Harik and Hotchkiss [KAP 01], and Wong and Law [WON 99] readers can discover technical details and in Kline [KLI 16] and Maruyama [MAR 98] they can find practical illustrations. And even when all these conditions are satisfied, it may be that the model is empirically under-identified. Multicollinearity may be the cause of this or, similarly, it may be a path coefficient that is close to zero and whose virtual absence affect condition rank of the concerned variables.

The possibilities that non-recursiveness offers must not lead us to forget the hidden risks, especially that of using it to shirk or camouflage theoretical weaknesses related to the direction of the predictive effects in the model that in the specified model. Given that in cases of theoretical doubt it is tempting to propose bidirectional relationships between variables, it would be useful not to give in to the ease of doing this.

You can find below the specification, in lavaan syntax, of the non-recursive model from Figure 3.4c:

STEP 2. Model specification (non-recursive, Figure 3.4c)

```
model.SPE <- 'Y1 ~ X1 + X2 + Y2
              Y2 ~ X2 + Y1
              Y3 ~ X1 + X2 + Y1 + Y2'
```

3.3.8. *Illustration of a path analysis model*

3.3.8.1. *A brief introduction to the theoretical model*

A simple model is presented below as an illustration. It is taken from a study carried out on the role of aging self-stereotypes (i.e., aging self-perceptions: the cognitions that an individual holds about him or herself as an aging person) among the elderly.

The authors of this study proposed testing a model that brought into play the relationships between psychological resources (self-esteem, dispositional optimism) and aging self-sterotypes and physical health among the elderly (N=331). This was a hypothetical model based in the socio-cognitive model of stereotypes and the conceptual framework of positive psychology, which postulates that positive emotions traits protect physical health A self-stereotype may be defined as a cognitive process through which individuals belonging to a social group tend to perceive themselves with positive and negative stereotypical traits and characteristics of the in-group. This may serve as a kind of cognitive filter that acts upon one's way of thinking, being and acting. Aging self-stereotypes are thus stereotypes integrated by the elderly, which were developed by society with respect to their social group. For example, in negative self-stereotyping, the elderly attribute terms such as "decline", "illness" and "senility" to themselves and see themselves as being a population "with problems" and that is "dependent".

As illustrated in Figure 3.5, this model hypothesizes the total mediation of aging self-sterotypes of age between psychological resources and objective (i.e. physical) health. It reflects the hypotheses according to which a person's psychological resources influence their aging self-stereotypes. Which in turn influence their physical health.

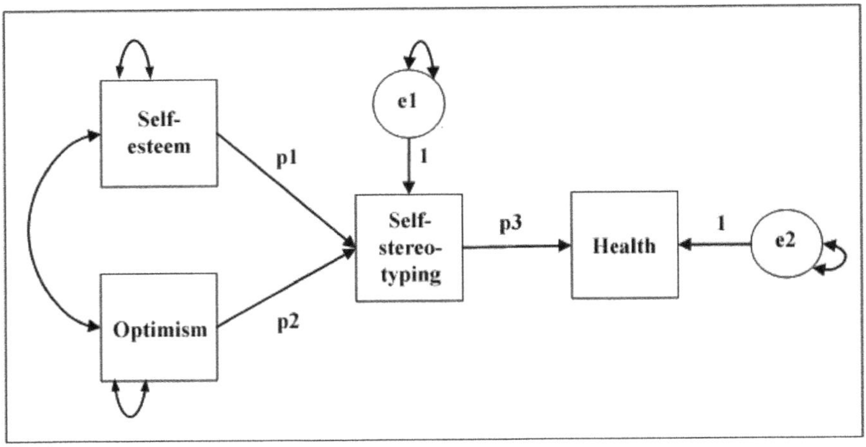

Figure 3.5. *Path model (the curved arrows with two head represent variances and covariances/correlations)*

We can see that the model calls upon four observed measured variables with two exogenous variables correlated with each other. These variables refer to two positive resources. The first was self-esteem, measured using Rosenberg's self-esteem scale [ROS 65], where a high score indicates high self-esteem; the second is dispositional optimism measured through the life orientation test [SCH 94], where a high score is indicative of a high level of optimism. Measured using two items, "physical health" is the ultimate endogenous variable, whose R^2 makes it possible to show the part of variation due to all variables that preceded it in the model. A high score is indicative of poor physical health. Between the two exogenous variables and the ultimate endogenous variable we can find the second, mediator endogenous variable related to aging self-stereotypes, measured through a five-item scale proposed by Levy, Slade, Kunkel *et al.* [LEV 02], where higher score indicates a high level of negative self-stereotypes, that is a negative perception of one's own aging.

This is a recursive model whose hypothesis of total mediation is reflected in the diagram by the fact that the exogenous variables have no direct influence on the ultimate endogenous variable (health). Their influence on this variable is mediated by the aging self-stereotypes (for more technical details on mediation, see [MAC 08]).

3.3.8.2. *Specifying the model in lavaan syntax*

The model represented in Figure 3.4 contains three causal paths, labeled "p1", "p2", and "p3". As we have seen, it has two endogenous variables, thus leading to two equations as well as an indirect effect leading to a specification. This is the product of the effects of the three causal path p1, p2 and p3, which will be notified to lavaan in order to obtain its value. Let us recall that by default lavaan correlates the exogenous variables of the model.

STEP 2. Model specification.

```
model.SPE <- 'stereotype ~ p1*esteem + p2*optimism
health ~ p3*stereotype
# Specify the indirect effect.
Indirect:= p1*p2*p3'
```

STEP 3. Using the model-fitting function "SEM".

```
model.EST <- sem (model.SPE, data = BASE)
```

STEP 4. Retrieving the results (solution) of the estimated model.

```
summary (model.EST, fit.measures = TRUE,
standardized = TRUE, modindices = TRUE, rsq = TRUE)
```

Upon studying model.SPE, we can see that the first equation translates the hypothesis that the variable "stereotype" is influenced by predictive effects from the variables "esteem" ("p1" is the label assigned to this parameter) and "optimism" ("p2" is the label assigned to this parameter). The second equation translates the hypothesis that the objective health "health" is subject to the predictive effect of the variable "stereotype" ("p3" is the label assigned to this parameter). It was necessary to use an asterisk to name the parameters (the direct effects) in order to calculate the indirect effect, which is the product of all direct effects (p1, p2 and p3). Here, we can specify the indirect effect of self-esteem on health (p1*p3) and/or that of optimism on health (p2*p3).

3.3.8.3. *Evaluation of the solution (model evaluation)*

The estimation of the model whose solution we will evaluate was done through the method of maximum likelihood, which is the default estimator in lavaan. We will examine the overall model fit, the quality of the solution (proper vs. improper solution) and the identification of possible zones of weaknesses, as well as the local fit indices of the solution.

The output of the solution is given in its minimal, summarized form in Table 3.1. We can see that these results can be broken down into two parts. The first displays overall goodness-of-fit indices, while the second offers local fit indices of the solution, including the option of the modification indices, if necessary. Before examining the contents of each in detail, we can observe that the solution has converged normally after 16 iterations. The solution seems to be proper.

3.3.8.3.1. The overall goodness-of-fit indices

Going by the statistically significant value of χ^2 (2, N = 331) = 6.48, p = 0.039, the harmony between the model and the data is not perfect. However, the sensitivity of this statistical test to sample size is too well-known to completely trust it. Indeed, the other fit indices, such as the CFI (0.971), the TLI (0.929) as well as the SRMR (0.036) argue in favor of the theoretical model subject to estimation. However, the rather small number of *df* (=2) could feed the implicit suspicion that this good fitting model could be due more to its lack of parsimony than to its underlying theoretical *substratum* that it represents. Indeed, the value of RMSEA (0.082), which contains the advantage of penalizing the lack of parsimony, supports this suspicion. The suspicion is further reinforced when we examine the higher limit of the confidence interval of the RMSEA whose value (0.157) goes beyond 0.100.

```
lavaan (0.5-23.1097) converged normally after  16 iterations

  Number of observations                           331

  Estimator                                         ML
  Minimum Function Test Statistic                6.485
  Degrees of freedom                                 2
  P-value (Chi-square)                           0.039

Model test baseline model:

  Minimum Function Test Statistic              162.194
  Degrees of freedom                                 5
  P-value                                        0.000

User model versus baseline model:

  Comparative Fit Index (CFI)                    0.971
  Tucker-Lewis Index (TLI)                       0.929

Loglikelihood and Information Criteria:

  Loglikelihood user model (H0)              -3157.764
  Loglikelihood unrestricted model (H1)      -3154.522

  Number of free parameters                          5
  Akaike (AIC)                                6325.528
  Bayesian (BIC)                              6344.539
  Sample-size adjusted Bayesian (BIC)         6328.679

Root Mean Square Error of Approximation:

  RMSEA                                          0.082
  90 Percent Confidence Interval          0.016  0.157
  P-value RMSEA <= 0.05                          0.165

Standardized Root Mean Square Residual:

  SRMR                                           0.036
```

Table 3.1a. *Goodness-of-fit indices for the mediation model (Figure 3.5)*

3.3.8.3.2. Local fit indices of the solution

A study of these indices, presented in the second part of the output (see Table 3.1b) makes it possible, first of all, to ensure that we get a proper solution. It can be verified (see the column "Estimate") that this contains no inadmissible parameter estimates (for example, in "Variance", a negative variance, called the *Heywood case*). And since the acceptability of the model is not doubted, we can proceed to the reading of its results.

Table 3.1b shows that the value of the three path coefficients have proven to be statistically significant. Let us specify here that the column "Estimate" gives the non-standardized coefficients, the column "P (>|z|)" gives the *p-value* and the column "Std. all" displays the standardized coefficients. We can thus note that the predictive effects of self-esteem (β = – 0.289, p = 0.000) and of optimism (β = – 0.355, p = 0.000) on aging self-stereotypes are negative, indicating that the higher the self-esteem and the more optimistic a person is, the less likely they will be to hold negative aging self-sterotypes related to aging. As for the effect of these self-stereotypes on physical health, it is positive (β = 0.353, p = 0.000), indicating that the more one holds negative aging self-perceptions, of age the poorer their physical health will be.

Although quite a weak effect [KLI 16], the indirect effect (mediation) of psychological resources on physical health has also proven to be statistically significant (0.036, p = 0.000).

On studying the squared multiple correlations (R^2, *R-square*) obtained for each endogenous variable, it can be noted that the share of variance of aging self-stereotypes that can be attributed to the set of predictive variables is almost 29% (R^2 = 0.286), while it is only around 13% as regards physical health (R^2 = 0.125).

Finally, the modification indices (the column "mi" in the spread shown in Table 3.1b) suggest two modifications in nature that could improve model fit: 1) modification 12 (Health ~ esteem), which introduce the effect of self-esteem on health and make it possible to reduce the χ^2 value by 5.782 points; 2) modification 15 (Health ~ esteem), which introduces the effect of health on self-esteem and makes it possible to reduce the χ^2 value by 6.406. The first modification is as plausible as the second is impossible. This is because an exogenous variable (esteem) cannot be subject to any predictive effect. Let us specify here that any modification introduces an additional parameter to be estimated in the model, thus reducing the degrees of freedom by one point. Nonetheless, it is essential that the decrease in the χ^2 value be $\geq$ 3.84 for a df = 1 in order that it may be considered as significant at $p < 0.05$, and ≥ 6.63 for a $df = 1$ in order that it may be considered significant at $p < 0.01$ (refer to a table for χ^2 values).

```
Parameter Estimates:

  Information                                 Expected
  Standard Errors                             Standard

Regressions:
                   Estimate  Std.Err  z-value  P(>|z|)   Std.lv  Std.all
  stereotype ~
    esteem     (p1)  -0.073    0.013   -5.773    0.000   -0.073   -0.289
    optimism   (p2)  -0.164    0.023   -7.077    0.000   -0.164   -0.355
  health ~
    stereotyp  (p3)   0.428    0.062    6.861    0.000    0.428    0.353

Variances:
                   Estimate  Std.Err  z-value  P(>|z|)   Std.lv  Std.all
   .stereotype        1.783    0.139   12.865    0.000    1.783    0.714
   .health            3.215    0.250   12.865    0.000    3.215    0.875

R-Square:
                   Estimate
    stereotype        0.286
    health            0.125

Defined Parameters:
                   Estimate  Std.Err  z-value  P(>|z|)   Std.lv  Std.all
    Indirect          0.005    0.001    4.351    0.000    0.005    0.036

Modification Indices:

          lhs op      rhs    mi    epc sepc.lv sepc.all sepc.nox
10 stereotype ~~   health 1.505 -0.302  -0.302   -0.100   -0.100
11 stereotype  ~   health 1.505 -0.094  -0.094   -0.114   -0.114
12     health  ~   esteem 5.782 -0.042  -0.042   -0.136   -0.022
13     health  ~ optimism 0.058  0.008   0.008    0.014    0.004
15     esteem  ~   health 6.406 -0.469  -0.469   -0.143   -0.143
18   optimism  ~   health 1.391  0.121   0.121    0.068    0.068
```

Table 3.1b. *Local indices of the solution (contd.)*

3.4. Actor-partner interdependence model

The model known as the *Actor-Partner Interdependence Model* (APIM) and popularized by Kenny, Kashy and Cook [KEN 06], is applied to dyadic data and is both non-recursive and just-identified (saturated). Figure 3.6 offers a graphic representation of this.

Non-recursive as it contains a correlation between the residual variables (e1 and e2), signifying the non-independence of the parts of the variance that are not explained by the variables in the model.

Just-identified as its df = zero. Indeed, there are as many variances-covariances (4*5/2 * 10) = 1, as there are parameters that are free to be estimated, namely four causal paths (a2, a1, p2) p2), two covariances, four variances (for a total of 10 parameters).

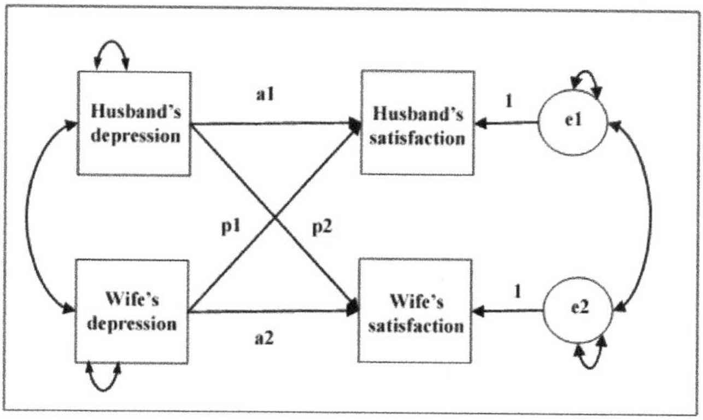

Figure 3.6. *APIM model (the curved double arrows represent the variances and covariances/correlations)*

What is expected from this model is not how well it fits the data, but the significance of the predictive effects that it hypothesizes. We can distinguish between two types of effects: the actor effects, which refer to intra-partner effects (for example, path "a1" refers to the effect that the husband's depression has on his own marital satisfaction) and partner effects, which are the inter-partner effects (for example, the path "p1" refers to the effect the husband's depression has on his wife's marital satisfaction).

If we wish to test the equivalence of these two effects within the dyad, simply constrain them to be equal them to be equal. Constraining paths a1 and a2 to be equal makes it possible to test the equivalence of intra-partner effects, while constraining p1 and p2 to be equal makes it possible to evaluate the equivalence of the inter-partner effects. Constraints of this kind make the model over-identified, with either a df=1, if the equality has a bearing on a single effect (intra *versus* inter), or a df=2 if the equality concerns all the effects (intra and inter). In both cases, a significant χ^2 indicates a significant difference between the effects. For example, the effect of a husband's depression on the marital satisfaction of his wife may be stronger than the effect of the wife's depression on the marital satisfaction of her husband.

Moreover, the APIM also depends on the type of the present dyad. According to Kenny, Kashy and Cook [KEN 06] there are two types of dyads: the dyad where both the two members are distinguishable (for example, a heterosexual couple) and a dyad where the two members are indistinguishable (for example, a homosexual couple). The analysis of indistinguishable dyads is related to intraclass correlations. Gonzalez and Griffin [GON 99] proposed a test to determine empirically if dyad members are distinguishable. Our illustration here focuses on dyads the two members are distinguishable.

3.4.1. *Specifying and estimating an APIM with lavaan*

We present three steps (model specification, model estimation, and retreiving the model solution, see Table 2.3) to test two APIMs: one that has no constraint for equality and the other with a constraint for equality of inter-partner effects.

The model establishes a dyadic relationship between marital satisfaction ("husbsatis", "wifesatis") and depression ("husbdep", "wifedep").

STEP 2. Specification of the APIM model (by default, lavaan correlates the measured exogenous variables).

```
model.SPE <- 'husbsatis ~ husbdep + wifedep
             wifesatis ~ wifedep + husbdep
            husbsatis ~~ wifesatis' # Correlation
between e1 and e2.
```

STEP 3. Estimating the APIM.

```
model.EST <- sem (model.SPE, data = BASE)
```

STEP 4. Retrieving the results of the APIM.

```
summary (model.EST, standardized = TRUE)
```

This syntax is used to translate the two equations of the model: 1) "husbsatis" is subject to the effets of "wifedep" and "husbdep"; 2) "wifesatis" is "subject to the effets of "wifedep" and "husbdep". We can note that by default lavaan will include the error terms "e1" and "e2" whose correlation was specified by using the double tilde (~~). Let us clarify here that when we specify a correlation between two endogenous variables (in this case, "husbsatis" and "wifesatis"), it is a correlation of their error terms that we specify as endogenous variables can never be correlated between themselves. Let us recall again that in SEM, only exogenous variables (whether manifest or latent) can be correlated between themselves, but this can never be the case with endogenous variables.

3.4.2. *Evaluation of the solution*

The solution presented in Table 3.2 is divided into two parts. The first encloses the first five lines indicating that convergence has occurred without any problem, that the sample contains 198 dyads (note that the couple is the unit of analysis), that the estimator, used is the maximum likelihood (ML) and, finally, that we are in the presence of a saturated model (dfχ = 0). The second part presents the estimation of the model's parameters, in this case, the four causal paths (reported in "regressions" section of the output) that are all statistic significant (see the column "p (>|z|)" and "Std.all" for the standardized estimations.

Indeed, it can be noted that the two actor effects are significant: the depression in the man negatively predicts his own marital satisfaction ($\beta = -0.426$) and, similarly, for the woman, her depression negatively predicts her own marital satisfaction ($\beta = -0.287$). As for partner effects: one person's depression negatively predicts the marital satisfaction of the other; the man's depression → marital satisfaction of the woman, $\beta = -0.252$; the woman's depression → marital satisfaction of the man, $\beta = -0.165$. We can also note that despite lavaan correlating the exogenous variables by default (that is, "husbdep", "wifedep") it still does not provide an automatic result for this. The displayed in "Covariances" section is that of the residual variables (36.578).

```
lavaan (0.5-23.1097) converged normally after  41 iterations

  Number of observations                           198

  Estimator                                         ML
  Minimum Function Test Statistic                0.000
  Degrees of freedom                                 0

Parameter Estimates:

  Information                                 Expected
  Standard Errors                             Standard

Regressions:
                   Estimate  Std.Err  z-value  P(>|z|)   Std.lv  Std.all
  husbsatis ~
    husbdep          -1.291    0.198   -6.507    0.000   -1.291   -0.426
    wifedep          -0.577    0.228   -2.524    0.012   -0.577   -0.165
  wifesatis ~
    wifedep          -1.109    0.264   -4.207    0.000   -1.109   -0.287
    husbdep          -0.844    0.229   -3.687    0.000   -0.844   -0.252

Covariances:
                   Estimate  Std.Err  z-value  P(>|z|)   Std.lv  Std.all
 .husbsatis ~~
   .wifesatis        36.578    5.241    6.979    0.000   36.578    0.571

Variances:
                   Estimate  Std.Err  z-value  P(>|z|)   Std.lv  Std.all
   .husbsatis        55.519    5.580    9.950    0.000   55.519    0.740
   .wifesatis        73.860    7.423    9.950    0.000   73.860    0.802
```

Table 3.2. *Lavaan output of the APIM*

One would be legitimate in now questioning the equivalence of the actor/partner effects within the dyads. For example, we would like to know whether the effect of the woman's depression on her spouse's marital satisfaction is stronger than the effect of the man's depression on his spouse's marital satisfaction. In order to find the answer, it is enough to constrain these two effects to be equal and to measure the impact this has on the model fit. If this equality leads to a deterioration in the model fit, it can be concluded that the two effects are different. Here is an illustration of this, using the APIM discussed above.

In order to constrain the parameters to be equal, their label must be specified when the model is specified and the desired constraint(s) should be added to the syntax at this stage.

STEP 2. Specifying and labeling parameters for the APIM model.

```
model.SPE<- 'husbsatis ~ a1*husbdep + p2*wifedep
             wifesatis ~ a2*wifedep + p1*husbdep
             husbsatis ~~ wifedep
              p1 == p2'         # Constraining two paths
 (effects) to be equal.
```

STEP 3. Estimating the APIM.

```
model.EST<- sem (model.SPE, data = BASE)
```

STEP 4. Retrieving the results of the APIM.

```
summary (model.EST, standardized = TRUE)
```

3.4.3. *Evaluating the APIM re-specified with equality constraints*

The equality constraint that was imposed (see p1 – (p2) in "Constraints" section of the output), allowed the model to go from saturation (*df*= 0) to over-identification, as is proven by the degrees of freedom, whose number is now 1.00, for a χ^2 value of 0.566 (see Table 3.3). Is this value enough to deteriorate the model fit? As presented above, referring to a χ^2 value the table shows us that for 1 df, a value of at least 3.84 is needed in order to hope for significance at $p < 0.05$. Thus, a χ^2 (1) = 0.566, p = 0.456, which is not significant at $p < 0,05$, thus indicates that the equality constraint did not significantly worsen the model fit; and hence the two effects can be considered as being similar (-0.204 and - 0.212). Moreover, the "Slack value" (in "Constraints" section of the output) which signifies the difference between p1 and p2, is zero (0.000).

```
lavaan (0.5-23.1097) converged normally after  38 iterations

  Number of observations                           198

  Estimator                                         ML
  Minimum Function Test Statistic                0.566
  Degrees of freedom                                 1
  P-value (Chi-square)                           0.452

Parameter Estimates:

  Information                                 Expected
  Standard Errors                             Standard

Regressions:
                   Estimate  Std.Err  z-value  P(>|z|)   Std.lv  Std.all
  husbsatis ~
    husbdep   (a1)   -1.202    0.159   -7.565    0.000   -1.202   -0.397
    wifedep   (p1)   -0.710    0.144   -4.923    0.000   -0.710   -0.204
  wifesatis ~
    wifedep   (a2)   -1.228    0.211   -5.819    0.000   -1.228   -0.318
    husbdep   (p2)   -0.710    0.144   -4.923    0.000   -0.710   -0.212

Covariances:
                   Estimate  Std.Err  z-value  P(>|z|)   Std.lv  Std.all
 .husbsatis ~~
   .wifesatis        36.702    5.253    6.986    0.000   36.702    0.572

Variances:
                   Estimate  Std.Err  z-value  P(>|z|)   Std.lv  Std.all
   .husbsatis        55.634    5.591    9.950    0.000   55.634    0.742
   .wifesatis        74.013    7.439    9.950    0.000   74.013    0.805

Constraints:
                                               |Slack|
    p1 - (p2)                                    0.000
```

Table 3.3. *Results of the APIM with equality constraints*

3.5. Models with latent variables (measurement models and structural models)

If the available data and the sample size allow, the researcher could covert the path model in Figure 3.5 into a model with latent variables. It would, thus, take the form presented in Figure 3.7, where it can be seen that each of the latent variables is measured by a certain number (here, arbitrarily fixed) of observed variables (indicators/items). These, as well as the endogenous latent variables, are accompanied by residual variables ("e", "E"). Similar to the graph in Figure 1.9 presented in Chapter 1, the graph in Figure 3.7 represents what may be called a "general structural equation model".

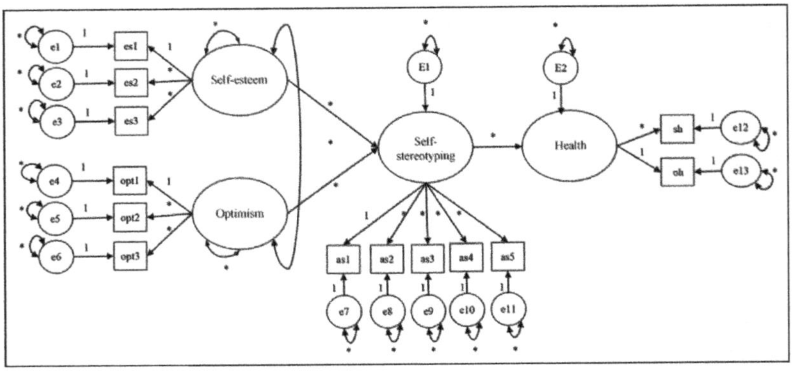

Figure 3.7. *General structural equation model with latent variables (the double curved arrows represent the variances and covariances/correlations; the asterisk denotes a free parameter to be estimated)*

Let us spend a little longer studying Figure 3.7. It is enough to break it down to grasp the underlying structure. This diagram is, in effect, made up of four connected parts. The following figures illustrate this breaking-down of the figure: Figure 3.8a represents the different measurement models that deal with the relationships between the observed variables and the latent variables, thus indicating how the former can be used to measure the latter. Figure 3.8b represents a structural model that only deals with the relationships between the latent variables, but it can also be applied exclusively to the measured variables (path models: refer to Figure 3.5). These two models are the basic components of structural equation modeling. They may be interconnected or may be used separately. Both these constituent components of SEM are fundamental and deserve to be discussed in greater detail.

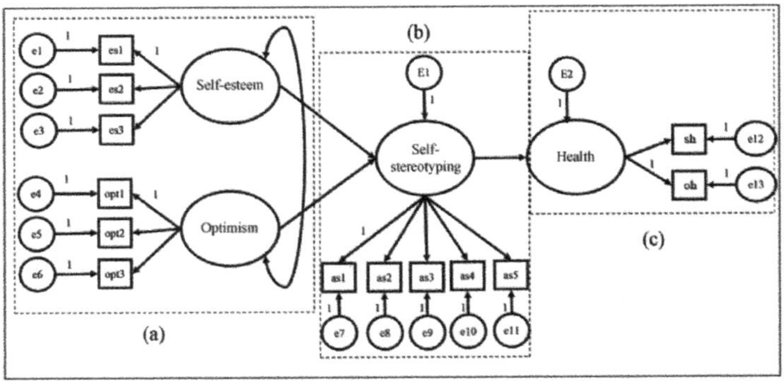

Figure 3.8a. *The three measurement models (a, b, and c) that make up the general structural equation model in Figure 3.7 (a = bidimensional model; b and c = unidimensional models)*

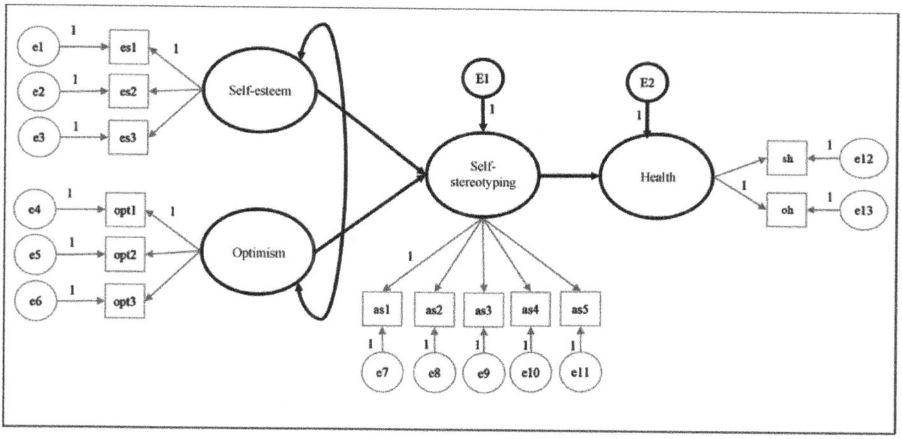

Figure 3.8b. *In bold, the structural portion of the model in Figure 3.7 (contd.)*

3.5.1. *The measurement model or Confirmatory Factor Analysis*

In SEM, a measurement model is equated to the Confirmatory Factor Analysis (CFA), which has been quite famous ever since Jöreskog [JÖR 73] introduced the first version of the LISREL software. Here, the confirmation covers the underlying structure of a set of indicators (for example, the items) selected to measure a construct. This involves testing hypotheses around a structure specified *a priori*. A beginner can refer to Mueller [MUE 96] for an accessible technical overview of CFA and can refer to Brown [BRO 15] for more practical examples of application.

The measurement model in used in SEM is based on the True Score Theory. The main objective of this theory is to evaluate score reliability. This refers to the precision with which the model is able to measure a construct. However, the measurement of a construct is never precise, never perfect, because it always carries a measurement error. Thus, the observed score (OS) obtained using a measurement instrument is never the true score (TS) – the true ability, which is unkown – but is a total measured score. That is, it is an additive composite of two components: true score/ability and measurement error (random error). Figure 3.9 presents this model whose equation is: Observed Score (OS) = True Score (TS) + Measurement Error (e).

According to this theory, every score for an indicator (i.e. for an item) is dependent both on the influences of the true score on the construct (for e.g. attributes) represented by the indicator, and on the measurement error (e). The reliability of a score increases as its measurement error approaches zero.

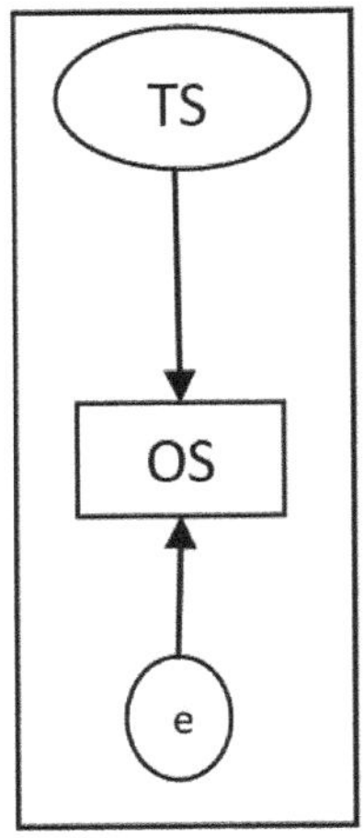

Figure 3.9. *Measurement model for the true score theory*

In other terms, the variance of an observed score is always equal to the sum of the variance of the true score and the variance of the measurement error. In reality, while the measurement error is made up of two additive parts, they are difficult to estimate separately. They are: random error (chance) and systematic error. At the empirical level, the sources of random errors are the various personal influences, such as the psychological state of the participant, their motivation, their sloppy reading of the items, their mood, or again, their concentration level when the test is administered. All these contribute to varying their responses on the proposed item. As concerns systematic errors, they refer to factors that are extrinsic to the participants, such as the measurement methods (self-reporting, hetero-evaluation) that can systematically introduce a bias in the responses of all respondents (for example, the "method effect")[5]. At the statistical level, a random error does not affect the mean score of a sample, but does affect the variability (distribution) of the scores around the mean. However, in the case of a systematic error it is quite different. This error seriously affects a population's mean score and biases the real perception of the phenomenon being studied. Such a bias compromises, for instance, the unidimensional nature of the construct whose item is the representative.

It is clear that these two kinds of error provide information on two different but complementary properties. One is precision (reliability), which is reflected in the amplitude of the random errors around the real score; the other is the accuracy of a measure (validity), which results from the absence of systematic errors in the

5 The reader will find an example of the method effect in [GAN 13].

conformity of the data with respect to the construct being considered. Accuracy gives information on the factor structure underlying the construct. This factor structure may be a unidimensional representation (congeneric), a multidimensional representation or, again, a hierarchic representation (second order, for instance) and a bifactorial representation (which is not the same as bidimensional).

3.5.1.1. *The unidimensional representation of a measure*

To illustrate this representation as well as the importance of CFA in the process of validating a measure, let us return to the scale used to measure life satisfaction that we discussed in the previous chapter.

Figure 3.10 illustrates the hypothesis of a unidimensional model. For the measured variables we have the five items that serve as indicators for a single latent variable representing the factor (F) "life satisfaction" (LS). When it is known that the orientation of an arrow indicates the direction of the effect of one variable on another, it then becomes easy to convert a graph that translates hypotheses around the relations between the variables in structural equations. A simple rule to be followed is: in a model there are as many structural equations as there are dependent variables (endogenous).

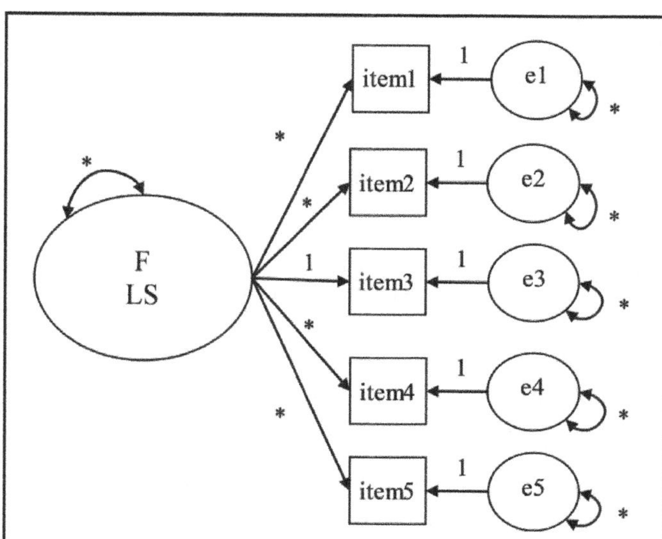

Figure 3.10. *Unidimensional representation of the Life Satisfaction (LS) scale (the curved double arrows represent variances and covariances; the asterisk designates a free parameter to be estimated*

An examination of the diagram in Figure 3.15 shows five dependent variables (in this case they are items). We can see that each item is subject to two influences, one being the common factor (F) and the other being a unique factor (measurement error (e)). We thus obtain the following equations:

item 1 = F + e_1

item 2 = F + e_2

item 3 = 1*F + e_3

item 4 = F + e_4

item 5 = F + e_5

The effect of the common factor reflects the factorial weight (B/β) and indicates the validity of the selected item. It is, thus, a construct validity coefficient. When squared, it represents that part of the variance that can be imputed to the common factor, that is, the construct. As concerns the effect of the unique factor, it represents the unique variance (uniqueness = an undistinguishable combination of specific factor and measurement error variance) in each item not explained by the common factor. The smaller the measurement error, the more reliable the item. In effect, when all the variance of an item is due to measurement error, this item cannot be a good indicator of the factor on which it is assumed to depend. It is clear that the smaller the measurement error, the more we tend towards factorial purity.

3.5.1.1.1. Identification of a measurement model

Among the required conditions for identifying a measurement model, there are two primordial conditions: having a sufficient number of indicators for the latent variable and defining the metric for this latent variable.

The first condition concerns the number of indicators per factor. Let us recall, here, that an indicator designates an item or again a measured/manifest variable. In order for a measurement model to be identified, it is important for the latent variable to have at least four indicators (for example: four items). In effect, with only three indicators, the measurement model is just-identified, thus making useless its testability.

The second condition relates to the metric for the latent variables. Because it is not measured, a latent variable clearly has no metric. It is thus up to the researcher to define the metric of this variable. There are two possible ways of doing this: one, they can set the variance of the latent variable to 1.00; otherwise, one parameter (i.e.; factor loading) relating the latent variable with an item can be set to 1.00, allowing this latent variable to then capture the metric of the selected item. This item will be used as a sort of reference indicator for scaling the latent variable. In effect, it is recommended that among the indicators designed to measure a construct, we choose the indicator that is the most representative, whose meaning approximates it the best. The object here is to offer readers a clear understanding of the construct being studied. For example, item 3, "I am satisfied with my life", could fulfill this purpose. Its perfect analogy with the construct of "life satisfaction" makes it a good reference indicator. By fixing the parameter of the reference indicator at 1.00 it can transmit its metric onto the latent variable. Following that a correspondence is set up between the two. Once the reference indicator is specified, a part of its variance will be transmitted to the latent variable for which it serves as the reference. We will examine the mathematical procedure in detail when we look at the results of the solution and, precisely, the variance of the latent variable.

What are the considerations that could guide the choice of these possibilities? It is clear that in the case of a CFA, using one or the other of these two options does not in any way affect the results of the analysis. On the other hand, it is not possible to fix the variance of the latent endogenous variables within a general structural equation model. To guarantee their metric, its imperative to set to 1.00 an observed measure/item on each latent variable as reference indicator. However, in both cases there may be theoretical and epistemological considerations that come in.

3.5.1.1.2. Model specification in lavaan syntax.

Let us recall that the "equal to" symbol followed by a tilde (=~) is the operator that is necessary for specifying a latent variable (see Table 2.2, Chapter 2). Let us also recall that the syntax specifying a model must be placed within simple, straight quotation marks (' ') and not the typographical quotation marks (‘ ’).

STEP 2. Specification of the measurement model in Figure 3.10.

```
model.SPE <- 'LS =~ item3 + item1 + item2 + item4 +
item5'
```

STEP 3. Model estimation using the model-fitting function "cfa" (*confirmatory factor analysis*) and robust estimator "MLR"

```
model.EST <-cfa (model.SPE, data = BASE, estimator =
"MLR")
```

STEP 4. Result retrieving

```
summary (model.EST, fit.measures = TRUE,
standardized = TRUE, rsq = TRUE, modindices = TRUE)
```

```
# Calculating the reliability coefficients (with semTools).

library(semTools)
reliability (model.EST)
```

Here, the above syntax translates the hypothetical model according to which LS (the arbitrary name given here) is a latent variable defined by the indicators: item1, item2, item3, item4 and item5. Let us recall that lavaan, by default, fixes at 1.00 the factor loading of the first item included in the equation at 1.00. Thus, to define the metric of the latent variable and identify the model, it is enough to place the reference indicator at the beginning of the equation, in this case item 3 ("I am satisfied with my life") of the life satisfaction scale (see Table 1.2, Chapter 1).

If we wish to identify the measurement model by setting the variance of the latent variable to 1.00, it is enough to free the constraint weighing upon the factor loading of the first item in the model specification:

```
model.SPE <- 'LS =~ NA*item3 + item1 + item2 + item4 + item5'
```

3.5.1.1.3. Model estimation

The model estimation syntax begins with indicating the fitting function "cfa" and continues with the choice of an appropriate estimator, which depends, as we have seen, on the sample size, the type of data and especially its multivariate normality. We will assess this using Mardia test (1970). In order to do this, we must use the package "semTools", which should be installed beforehand. This must then be opened and used to calculate the Mardia test. Here is an illustration, using the responses of 137 participants (N = 137) to five items of the life satisfaction scale.

```
> head (BASE)
  item1 item2 item3 item4 item5
1     4     4     4     4     4
2     5     4     4     2     4
3     3     4     4     2     2
4     4     3     4     2     2
5     4     4     4     4     5
6     3     4     3     3     3
> library (semTools)
> mardia<-mardiaKurtosis (BASE)
> round (mardia, 3)
   b2d      z      p
49.058  9.833  0.000
```

Table 3.4. *The script and results of the Mardia test*

After having imported the data called "BASE" (an overview of this can be seen by using the command "head (BASE)"), and after having launched "semTools" (library (semTools)) we run the calculation of the Mardia test "mardiaKurtosis (BASE)". In order to simplify the reading of the results and avoid the exponential notations found here, the number of digits after the decimal point was fixed to 3, "round (MARDIA, 3)".

The last line of the output (Table 3.4) provides the Mardia coefficients: the value 40,058 refers to the multivariate kurtosis coefficient (kurtosis = b2d), while the value 9.833 refers to the standardized multivariate kurtosis coefficient (*z*-score) accompanied by its p-value (*p*). It can be noted that not only is this standardized coefficient statistically significant, but its value is greater than 5, which is the threshold value beyond which multivariate normality seems to fail [KLI 16].

This observation suggests to us that we could use a different estimator from the default one, namely, maximum likelihood (ML). We chose a "robust" estimator, the MLR, as it seemed more appropriate to data obtained through an ordinal scale with five categories (see Table 1.10), using a fairly small sample size (N = 137). Let us now look at the solution obtained using this estimator.

3.5.1.1.4. Evaluation of the solution

Let us recall that the solution is examined with respect to three criteria: 1) the overall goodness-of-fit model; 2) the quality of the solution and the location of areas of potential weakness and misfit; and 3) the local fit indices of the solution (parameter estimates). The output of the solution, whose details we will comment on, are presented in the following minimal form (Table 3.5):

It can be noted that these results are divided into two parts. The first part displays the overall goodness-of-fit indices, while the second offers the local fit indices, including, if necessary, the modification indices. Before examining the respective content in detail, we can remark that the solution converged normally after 22 iterations, resulting in a proper solution.

```
lavaan (0.5-23.1097) converged normally after  22 iterations

  Number of observations                           137

  Estimator                                         ML       Robust
  Minimum Function Test Statistic                9.432        6.361
  Degrees of freedom                                 5            5
  P-value (Chi-square)                           0.093        0.273
  Scaling correction factor                                   1.483
    for the Yuan-Bentler correction

Model test baseline model:

  Minimum Function Test Statistic              140.548       84.441
  Degrees of freedom                                10           10
  P-value                                        0.000        0.000

User model versus baseline model:

  Comparative Fit Index (CFI)                    0.966        0.982
  Tucker-Lewis Index (TLI)                       0.932        0.963

  Robust Comparative Fit Index (CFI)                          0.984
  Robust Tucker-Lewis Index (TLI)                             0.967

Loglikelihood and Information Criteria:

  Loglikelihood user model (H0)               -888.426     -888.426
  Scaling correction factor                                   1.332
    for the MLR correction
  Loglikelihood unrestricted model (H1)       -883.710     -883.710
  Scaling correction factor                                   1.370
    for the MLR correction

  Number of free parameters                         15           15
  Akaike (AIC)                                1806.852     1806.852
  Bayesian (BIC)                              1850.652     1850.652
  Sample-size adjusted Bayesian (BIC)         1803.198     1803.198

Root Mean Square Error of Approximation:

  RMSEA                                          0.080        0.045
  90 Percent Confidence Interval          0.000  0.159        0.000  0.118
  P-value RMSEA <= 0.05                          0.217        0.477

  Robust RMSEA                                                0.054
  90 Percent Confidence Interval                              0.000  0.162

Standardized Root Mean Square Residual:

  SRMR                                           0.036        0.036
```

Table 3.5a. *Goodness-of-fit indices of the measurement model represented by Figure 3.10*

```
Parameter Estimates:

  Information                                      Observed
  Standard Errors                        Robust.huber.white

Latent Variables:
                   Estimate  Std.Err  z-value  P(>|z|)   Std.lv  Std.all
  LS =~
    item3             1.000                               0.638    0.712
    item1             0.914    0.125    7.322    0.000    0.583    0.626
    item2             0.672    0.173    3.879    0.000    0.428    0.507
    item4             0.858    0.224    3.833    0.000    0.547    0.558
    item5             1.245    0.243    5.120    0.000    0.794    0.625

Intercepts:
                   Estimate  Std.Err  z-value  P(>|z|)   Std.lv  Std.all
   .item3             3.825    0.076   50.008    0.000    3.825    4.272
   .item1             3.737    0.079   47.026    0.000    3.737    4.018
   .item2             3.847    0.072   53.295    0.000    3.847    4.553
   .item4             3.715    0.084   44.320    0.000    3.715    3.787
   .item5             3.409    0.109   31.406    0.000    3.409    2.683
    LS                0.000                               0.000    0.000

Variances:
                   Estimate  Std.Err  z-value  P(>|z|)   Std.lv  Std.all
   .item3             0.395    0.088    4.503    0.000    0.395    0.493
   .item1             0.526    0.121    4.343    0.000    0.526    0.608
   .item2             0.530    0.104    5.088    0.000    0.530    0.743
   .item4             0.663    0.139    4.777    0.000    0.663    0.689
   .item5             0.983    0.201    4.881    0.000    0.983    0.609
    LS                0.407    0.101    4.031    0.000    1.000    1.000

R-Square:
                   Estimate
    item3             0.507
    item1             0.392
    item2             0.257
    item4             0.311
    item5             0.391

Modification Indices:

     lhs op   rhs    mi mi.scaled    epc sepc.lv sepc.all sepc.nox
18 item3 ~~ item1 5.149     3.472  0.157   0.157    0.189    0.189
19 item3 ~~ item2 2.790     1.882 -0.096  -0.096   -0.127   -0.127
20 item3 ~~ item4 0.028     0.019 -0.012  -0.012   -0.013   -0.013
21 item3 ~~ item5 0.390     0.263 -0.059  -0.059   -0.052   -0.052
22 item1 ~~ item2 1.195     0.806  0.063   0.063    0.080    0.080
23 item1 ~~ item4 4.502     3.036 -0.143  -0.143   -0.157   -0.157
24 item1 ~~ item5 1.814     1.223 -0.122  -0.122   -0.103   -0.103
25 item2 ~~ item4 0.251     0.169  0.030   0.030    0.037    0.037
26 item2 ~~ item5 0.089     0.060  0.024   0.024    0.022    0.022
27 item4 ~~ item5 3.595     2.424  0.175   0.175    0.140    0.140
```

Table 3.5b. *Parameter estimates of the of the measurement model in Figure 3.10 (contd.)*

Overall goodness-of-fit indices

As we can see, lavaan provides goodness-of-fit indices generated by ML estimator as well as MLR estimator under the column, "Robust", using the Yuan-Bentier correction. We will look more specifically at the indices that appear in this last column.

Absolute Fit Indices (χ^2 and SRMR)

We can first observe the difference in the χ^2 values (*Minimum Function Test Statistic*), generated for these two different methods: $\chi^2_{ML}(5) = 9.432$ *versus* $\chi^2_{Robust}(5) = 6.361$ (=9.432/1.438, see equation [1.18], Chapter 1 of this book). The *scaling correction factor for the Yuan-Bentler correction,* equal to 1.483 (= 9.432/6,361) indicates a moderate correction, knowing that a value > 1.00 is indicative of a deviation from normality. It can be noted that the two $\chi^2(5)$, each accompanied by 5 df, are not statistically significant ($p > 0.05$), which suggests that there is excellent harmony between the model and the data. It is not surprising that the value of the SRMR (0.036) argues in favor of this harmony.

Finally, let us recall, that the *baseline model* refers to the zero model, also called the "independence model" as it hypothesizes the total independence of the items (Figure 1.6b, Chapter 1). Such a model is dedicated to total inadequacy.

The incremental fit indices (CFI and TLI)

In Chapter 1, we saw that the CFI and TLI used, as basis of their comparison, the χ^2 of the null model (the baseline model) which hypothesizes, improbably, that the variables of the model have no relationship with each other. It is not surprising that the χ^2 of this model is very high ($\chi^2_{ML} = 140.54$; $\chi^2_{robust} = 84.44$), raising doubts about the likelihood of the null model. It is also not surprising that the two indices argue in favor of the same conclusion as concerns how well the model fits the data. In effect, the CFI and the TLI display "robust", corrected values, greater than 0.95 (robust-CFI = 0.984 and robust-TLI = 0.967). Let us recall here that a model fit is all the better then the values of the CFI and TLI are close to 1.00, with a threshold value of acceptability of at least 0.95.

The parsimonious fit indices (RMSEA)

The corrected value of the RMSEA (robust-RMSEA, that is, 0.054, with a 90% confidence interval between 0.000 and 0.162, does not exceed the threshold value of acceptability (0.06), thus arguing that there is compatibility between the model and the data. We observe, however, that the upper limit of the confidence interval exceeds 0.1, thus penalizing a certain lack of parsimony, evidenced by the small number of degrees of freedom (=5) of the model.

It can, thus, be concluded that the overall goodness-of-fit indices argue in favor of a very good fit of the model to the data. While these results do away with the need to try and locate the weak points of the model by turning to the fit indices, they cannot dispense with the examination of local indices.

The local fit indices of the solution

This part of the results is divided into three sub-parts: the parameter estimates of the model ("Parameter estimates"), the R^2 ("R-squares") of the indicators and the suggested modifications with the aim of possibly improving the model fit ("Modification indices").

On consulting the "Parameter estimates" we can already see that the obtained solution contains no impurities such as inadmissible and offending estimates (negative variances) or very high/low standard error ("std.err"). The "estimate" column offers the non-standardized factor loading coefficients, each accompanied by its standard error ("std.err"). The direction of the coefficient (negative *versus* positive) indicates the nature of the predicted relationship between the indicator and the latent variable upon which it depends. For example, the non-standardized factor loading of item 1, that is, 0.914, indicates that an increase of one unit in the latent variable, "life satisfaction" is associated with an increase of 0.914 in item 1 (as a measured variable). The "z-value" column, whose values are obtained by dividing the "Estimate" by its "std.err" makes it possible to judge the statistical significance of the factor loading. For example, the *z*-value of the factor loading of the item 1 is equal to 7.322 (i.e. 0.914/0.125 = 7.32). A ratio with an absolute value greater than 1.96 signifies that the factor loading is significant (i.e. non-zero) at $p < 0.05$. It can be observed that all five factorial factor loading are significant as their statistics are greater than 1.96.

The standardized factor loading occupy the last two columns, called "std.lv" and "std.all". The "std.lv" column presents the factor loading of a solution where only the latent variables were standardized (i.e. $M = 0.00$, $ET = 1.00$) while the "std.all" column presents the factor loading of a solution where again the latent variables and indicators were standardized (completely standardized solution). This is the option that must be studied and evaluated. We will see that, for example, all the factor loading are positive and greater than 0.40, which is the required minimum to judge the relevance of the link between an item and the factor on which it depends.

However, as we will see further down, this required minimum is not unanimously accepted.

It is also possible to obtain a 95% CI for the parameter estimates. In order to do this, simply use the following function:

```
parameterEstimates (model.EST)
```

That part of the results that is relative to variances deserves to be paid attention, especially as regards the variance of the latent variable (LS), which is equal to 0.407. We had written, earlier, that one part of the variance of the reference indicator is transferred to the latent variable for which it serves as the reference. The variance of item 3, the reference indicator in our model, is 0.807. Its factor loading, standardized by the latent factor, is 0.712. When squared ($0.712^2 = 0.50$), this factor loading indicates what percentage of variance (50%) of the item can be imputed to the factor on which it depends. Thus, 50% of the variance of item 3 (which is 0.807) is transferred to the latent variable (i.e. 50% of 0.807).

The variance of the latent variable VAR (LV) λri is obtained as follows (with as an illustration of the life satisfaction scale measurement model):

$$\text{VAR (LV)} = (\lambda_{ri})^2 \, \sigma_{ri} \quad [3.6]$$

$$= (0.712)^2 \, 0.807$$

$$= (0.506) \, 0.807$$

$$= 0.408$$

where:

– λ_{ri} refers to the standardized factor loading of the reference indicator (ri);

– σ_{ri} refers to the variance of this indicator.

The variance of the indicator (σ_{ri}), which, like the standard-deviation, is calculated from the raw data, must not be confused with the error variance of the indicator that figures in the results table under the heading “variances”.

It should be noted that the value 0,408 corresponds to the 0,407 shown in the results at the intersection of the "estimate" column and the "variance" line of LS (Table 3.5b).

As concerns the R^2 ("R-square"), it gives a great deal of information on the quality of the indicator as a representative of the latent variable on which it depends.

An R^2 is obtained by squaring the completely standardized factor loading coefficient. The true variance of each item can be obtained by squaring its coefficient of factor loading (λ^2). This is the fraction of variance explained by the common factor. The other part of the total variance (equal to 1 - λ^2) is due to the error. For example, a standardized value of $\lambda^2 = 0.39$, signifies that 39% of the variance of the concerned item is determined by the common factor, and the rest (1-0.39 - 0.61) or 61%, is determined by the unique factor (for example, measurement errors). These factorial factor loading coefficients express the degree of convergence between the latent variables and their indicators. They make it possible to estimate the quality of the relationship between an item and the factor on which it depends (i.e., validity coefficient). We can see that the highest R^2 is related to item 3, considered to be the reference item as it explicitly refers to life satisfaction ("I am satisfied with my life"). Some psychometricians (for example, [COM 92, TAB 07]) recommend that we choose items whose R^2 is greater than 0.40 (thus showing a factorial factor loading greater than 0.63).

The final part of the results provides the modification indices, which deserve to be clarified. These indices can be useful when the model's fit is problematic. The "rhs" (*right hand* side), "op" (*operator*) and "lhs" (*left hand side*) columns all present variable pairs (for example, "item3" and "item1") and the nature of the suggested relationships between them, through the operator (for example, "~~" for *correlation*). The introduction, through copy-paste, of such a suggestion within the model (here: item3 ~~item1) to correlate error variances of item 3 and item 1 could improve model fit. The consequences of such a modification on the χ^2 of the model are given first in the columns "mi" (for the χ^2_{ML}) and "mi-scaled" (for the χ^2_{robust}). We will examine this later more closely. Freeing the correlation of these two error variances brings down the χ^2_{robust} by 3.472 points. Is this reduction enough to allow an improvement of model fit? No. This is because a decrease of at least 3.84 points is required, reflecting the critical value of a χ^2 for 1 *df* at $p < 0.05$ (simply consult a χ^2 table). The *df* comes from the fact that we have freed the correlation between the two two error variances which becomes a new parameter that is free to be estimated and thus reduces the *df* of the original model by one point. Each modification lowers the *df* of the model by one point. And, in order for such a modification to claim to improve the the overall model fit, the consequent decrease in value of χ^2 must be greater than 3.84. It can be seen, for example, by examining the "mid-scaled" column, that no decrease reaches the critical value 3.84, and therefore none is able to improve the model's fit. The column "epc" (*expected parameter change*) gives the non-standardized value of the new parameter if it was specified in the model. The last columns give the standardized value based on the type of standardization. For example, "std.nox" offers a solution where all variables, apart from the exogenous variables, have been standardized.

Finally, a brief note about the section on intercepts, which owes its presence only to having used the MLR estimator. In principle, the intercepts are displayed when the command "meastructures = TRUE" is activated in the model estimation where the analysis covers the variances-covariances as well as the means of the indicators (for example, multigroup analyses, longitudinal studies). The intercept of an indicator thus refers to its estimated mean when the factor it depends on is equal to zero. We will return to this further in the text.

3.5.1.1.5. Matrix of observed (S), reproduced (Σ) and residual (S – Σ) covariances/correlations

The reader can easily replicate Table 1.5, presented in Chapter 1 of this book. The matrix of observed covariances/correlations (S) in our sample is easy to obtain, using the following commands:

```
# To obtain the matrix of observed correlations (S).

cor (BASE)

# To obtain the matrix of observed covariances (S).

cov (BASE)
```

As concerns the model-implied covariance matrix (Σ), which is the very essence of structural equation modeling, the reader can either manually calculate it or obtain it through lavaan using some simple commands.

Let us first examine manual calculations. The simplicity of our model lends itself to this since it contains only ten covariances/correlations. The product of two factor loadings factorial “std.all” and “std.lv” of two items reproduces their correlation and covariance respectively. To clarify, let us take item 3 and item 1 (Table 3.5b). Their reproduced correlation is equal to (0.712) (0.626), that is 0.445 (0.45, rounding off), with their observed correlation being 0.505. Their reproduced covariance is equal to (0.638) (0.583), that is, 0.371, while their observed covariance is 0.424.

By proceeding in this way, the reader will be able to reproduce both the covariance matrix and the correlation matrix (Σ). The differences between the two matrices S and Σ generate the residual covariance/correlation matrix which is obtained by substracting the value of element of the reproduced matrix from that of the corresponding element in the observed matrix (S – Σ). The residual correlation of item 3 and item 1 is, thus, equal to 0.505 – 0.445, or 0.06.

The reader can compare the results obtained manually with those obtained through lavaan, by using the following commands:

```
fitted (model.EST)     # To obtain the model-implied covariance matrix (Σ).
```

In order to convert the covariances matrix into a correlations matrix, simply proceed as described below:

```
cov.reproduced <- fitted (model.EST)$cov     #cov.reproduced is an arbitrary name.
cor.reproduced <- cov2cor (cov.reproduced) #cov.reproduced is an arbitrary name.
cor.reproduced     # For the matrix (Σ)
```

Finally, in order to obtain the residual matrix (S – Σ) one of the following commands is required:

```
# To obtain the residual correlations matrix (S - Σ).

residuals (model.EST, type = "cor")

# To obtain the residual covariances matrix (S - Σ).

residuals (model.EST, type = "raw")
```

3.5.1.1.6. Reliability coefficients

The solution to a CFA is not limited only to the overall fit indices, which, through the evaluation of the conformity of model fit to data, seeks in fact to test its structural validity. It also makes it possible to analytically estimate the reliability of each item through its R^2 and, overall, the reliability and the internal consistency of a measure made up of all the items.

As concerns the reliability of a group of items with respect to a given factor, this can be estimated using several coefficients, the best known of which is Cronbach's alpha [CRO 51], based on the correlations/covariances between items. However, it is also possible to calculate others using the results of the CFA (i.e. the factorial factor loadings). For example, the composite reliability coefficient also called "omega") proposed by Raykov [RAY 01] can replace Cronbach's alpha:

$$\omega = \frac{(\Sigma\lambda_i)^2}{(\Sigma\lambda_i)^2 + \Sigma(\theta\delta_i)} \qquad [3.7]$$

where:

– λ_i = non-standardized estimate (loading) of the item "i" by the latent factor;

– $\theta_{\delta i}$ = the error variance for the item "i".

This is, in fact, a composite index that evaluates the internal consistency of a measure. It indicates the degree of homogeneity of the content of the items used to measure the theoretical construct (the factor). The value of this index varies between 0 and 1. The closer this value is to 1, the greater the reliability. This formula shows how the definition of internal consistency is entirely subordinate to the error variance of the items. In addition, it should be emphasized that a high alpha coefficient is not an indicator of the unidimensionality of a measure [COR 93].

lavaan calculates four coefficients, including the Raykov omega coefficient, using the "reliability ( )" function. Here is an illustration applied to the unidimensional model of the life satisfaction scale:

STEP 2. Specification of the measurement model in Figure 3.10.

```
model.SPE <- 'LS =~ item3 + item1 + item2 + item4 +
item5'
```

STEP 3. Model estimation using the model-fitting function "cfa"

```
model.EST <- cfa (model.SPE, data = BASE)
```

STEP 4. Retrieving of the results including the modification indices

```
summary (model.EST, fit.measures = TRUE,
standardized = TRUE, modindices = TRUE)

# Calculating the reliability coefficients (with semTools).

reliability (model.EST)
```

Table 3.6 shows the command and the resulting output.

```
> reliability (model.EST)
               LS     total
alpha  0.7367880 0.7367880
omega  0.7427161 0.7427161
omega2 0.7427161 0.7427161
omega3 0.7405397 0.7405397
avevar 0.3751853 0.3751853
```

Table 3.6. *Reliability indices*

It can be noted that the values of the four output reliability coefficients (i.e. Cronbach's alpha [CRO 51], Raykov's omega [RAY 01a], Bentler's omega 2 [BEN 09] and McDonald's omega 3 [MCD 99]) are close to each other. The value of these indices range from 0 (zero reliability) to 1 (excellent reliability). Reliability increases the closer the value gets to 1.00, with an acceptability threshold of 0.70. Let us also recall that these indices are sensitive to the number of items. The last line shows a different value. This is "avervar", the value of the Average variance extracted (AVE; [FOR 81]) for each latent variable in the model.

3.5.1.2. *Bi/multidimensional representation of a measure*

This representation refers to oblique factor solutions. To illustrate, let us look at the dispositional optimism scale developed by Snyder and colleagues [SNY 91]. This scale comprises of eight items representing two correlated dimensions: *Pathways* (four items: p1, p3, p4, p5) and *Agency* (four items: a2, a6, a7, a8). The graph in Figure 3.11 translates the hypothesis of this bidimensional model: two common factors affecting each of the items specified in advance and measuring two different, but connected, aspects of the construct. We can see that four items depend only on the "Pathways" factor and their respective unique factor. The four other items depend only on the "Agency" factor as well as on their respective unique factor. While the possibility is not excluded, it is rarely possible for an item to depend on several factors at once (i.e., cross-loading). We can also note that measurement errors, here, are assumed to have no intercorrelations. However, we also cannot exclude the possibility that they could be intercorrelated. Finally, it is hypothesized that there is a covariation between two common factors; they are, in effect, connected on the diagram by a curved double arrow. When we parameterize the model, the covariance between these two factors will be considered as a free parameter to be estimated. In lavaan, the correlations between latent exogenous factors (dimensions) are estimated by default.

We can thus count eight structural equations, accompanied, this time, by a covariance between the two factors.

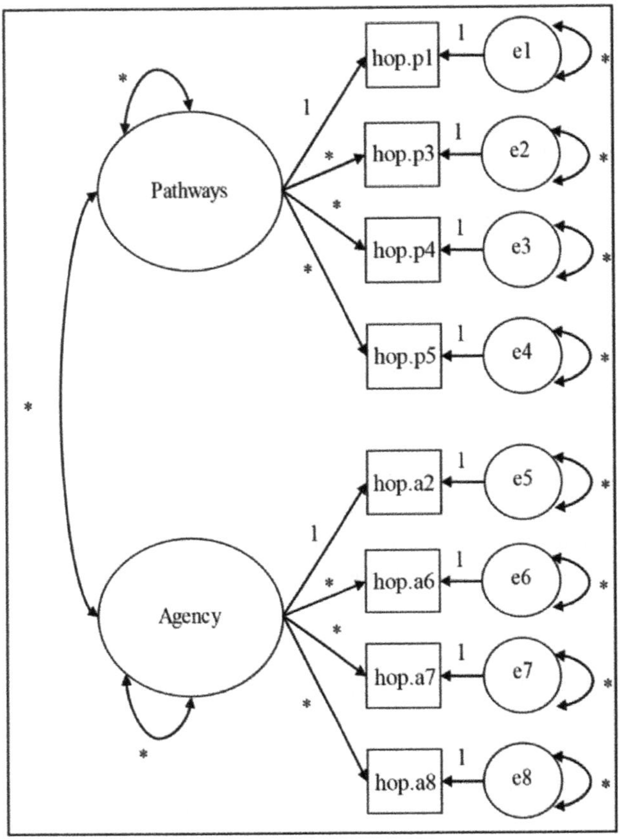

hop.p1 = pathways + e1

hop.p3 = pathways + e2

hop.p4 = pathways + e3

hop.p5 = pathways + e4

hop.a2 = agency + e2

hop.a6 = agency + e6

hop.a7 = agency + e7

hop.a8 = agency + e8

Figure 3.11. *Bidimensional model of the dispositional optimism scale and its equations (the curved double arrows represent the variances and covariances/correlations; the asterisk denotes a free parameter to be estimated)*

3.5.1.2.1. Model specification in lavaan syntax

It can be noted that "hop.p1", "hop.p3", "hop.p4", "hop.p5" are the names of the indicators of the dimension *Pathways*, as they appear in our file (BASE). We could have named them "item1", "item2", "item3", and "item4" for instance. We can also note that by default lavaan correlates latent exogenous variables. It is, thus, futile to add the following indication "pathways ~~ agency" when specifying the model. On the other hand, if we wish to remove this correlation from the model, it must be specified as follows: "pathways ~~ 0*agency".

STEP 2. Specification of the bidimensional model of hope scale (Figure 3.11), correlation between the dimensions estimated by default.

```
model.SPE <- 'pathways =~ hop.p1 + hop.p3 + hop.p4 +
hop.p5 agency =~ hop.a2 + hop.a6 + hop.a7 + hop.a8'
```

STEP 3. Model estimation using the model-fitting function "cfa"

```
model.EST <- cfa (model.SPE, data = BASE,
estimator = "MLR")
```

STEP 4. Retrieving the results including the modification indices.

```
summary (model.EST, fit.measures = TRUE,
standardized = TRUE, modindices = TRUE)
```

It can be noted here that the chosen estimation method is a method said to be "robust" (MLR) as the Mardia calculation coefficient (100.77, $p < 0.001$) indicates a severe violation of the multivariate normality of our data.

3.5.1.2.2. Evaluation of the solution

The only difference with respect to the solution of the unidimensional model (see the solution of the life satisfaction scale Table 3.5b) is found in the presence of a section dedicated to the covariances/correlations between the latent variables present in the model (Table 3.7). This aspect is far from anodyne, as it gives information on the discriminant validity of the variables present. The correlations between the latent variables in a measurement model must be interpreted according to the theoretical considerations underlying the model. For example, low correlation or the absence of any correlation between latent variables argues for good discriminant validity between the variables. On the other hand, high correlation, that goes beyond 0.80 or

0.85, brings into doubt this discriminant validity. We are rather in the presence of redundant or similar factors that it would be wiser to combine. It can be noted that the correlation between the two dimensions of the dispositional hope scale ("Pathways" and "Agency") is at 0.763, indicating that its two components are different from one another while being strongly connected to one another (see Table 3.7). Moreover, to be persuaded of this, it is enough to simply compare the fit of this bidimensional solution ($\chi^2 = 30.67$, $df = 19$) with that of a unidimensional solution ($\chi^2 = 54.13$, $df = 20$), which is not reported here.

```
Parameter Estimates:

  Information                                  Expected
  Standard Errors                              Standard

Latent Variables:
                   Estimate  Std.Err  z-value  P(>|z|)   Std.lv  Std.all
  pathways =~
    hop.p1            1.000                               0.858    0.641
    hop.p3            0.692    0.111    6.229    0.000    0.594    0.428
    hop.p4            0.940    0.113    8.344    0.000    0.806    0.615
    hop.p5            1.166    0.125    9.303    0.000    1.000    0.757
  agency =~
    hop.a2            1.000                               0.787    0.676
    hop.a6            1.116    0.157    7.084    0.000    0.878    0.515
    hop.a7            0.868    0.126    6.906    0.000    0.683    0.499
    hop.a8            0.888    0.111    8.036    0.000    0.699    0.613

Covariances:
                   Estimate  Std.Err  z-value  P(>|z|)   Std.lv  Std.all
  pathways ~~
    agency            0.515    0.079    6.522    0.000    0.763    0.763

Variances:
                   Estimate  Std.Err  z-value  P(>|z|)   Std.lv  Std.all
   .hop.p1            1.054    0.107    9.862    0.000    1.054    0.589
   .hop.p3            1.570    0.135   11.646    0.000    1.570    0.817
   .hop.p4            1.068    0.105   10.209    0.000    1.068    0.622
   .hop.p5            0.745    0.099    7.500    0.000    0.745    0.427
   .hop.a2            0.737    0.085    8.682    0.000    0.737    0.543
   .hop.a6            2.138    0.197   10.866    0.000    2.138    0.735
   .hop.a7            1.412    0.128   10.999    0.000    1.412    0.751
   .hop.a8            0.813    0.083    9.770    0.000    0.813    0.625
    pathways          0.736    0.133    5.524    0.000    1.000    1.000
    agency            0.619    0.110    5.638    0.000    1.000    1.000
```

Table 3.7. *Parameter estimates from the two-factor CFA model of hope*

3.5.1.2.3. Reliability coefficients

Here, the command that was already presented earlier ("reliability (model.EST)") makes it possible to obtain the four reliability coefficients (i.e. alpha, omega, omega

2, omega 3) for each factor (dimension) in the specified model. In the case of our example these would be "Pathways" and "Agency". Generally speaking, when using this command there can be as many results as there are dimensions (factors) in the hypothetical model specified and estimated.

3.5.1.3. *Hierarchical representation of a higher-order structure of a measurement model*

Figure 3.12 shows the hypothesis of a hierarchical factor solution (i.e., higher-order or second-order measurement model) in which "physical ability", "physical appearance", "relations with peers" and "relations with parents" are four first-order or lower-order factors, each influencing a group of items, but being, in turn, influenced by SDQ, a common second-order or higher-order factor. This is, in a way, the factor of factors. This is supposed to explain the relations between the first-order factors.

It is important to note that "physical ability", "physical appearance", "peers" and "parents" become, here, dependent (latent) variables whose variances and covariances are no longer estimated as they are assumed to depend on a second-order factor (SDQ). This is reflected in the diagram by the absence of curved arrows between "physical ability", "appearance", "peers" and "parents", and by the presence of arrows directed from SDQ towards "physical ability", "appearance", "peers" and "parents". Moreover, the higher-order factor loadings of "physical ability", "appearance", "peers" and "parents" from SDQ cannot be made without errors, hence the residual variables (E1, E2, E3, E4) associated with each of the the four lower-order factors that make it possible to infer the share of variance imputable to a higher-order factor and the portion that can be imputed to all that is extrinsic to the model (i.e. disturbance).

It must be emphasized that in order to be identified, such a model requires the presence of at least the four lower-order factors with factors because with only three lower-order factors the structural portion (i.e., higher-order factor loadings) is just-identified.

Finally, when testing the a hierarchical factor model, the following general sequence must be observed: 1) ensuring that the first-order model fits the data well (a multidimensional CFA must be applied, bringing into relation the lower-order factors of the model); 2) reviewing the oblique multidimensional solution and, above all, examining the amplitude and the direction of the correlations between the factors (for example, the very modest correlation between the first-order dimensions do not argue in favor of the likelihood of a second-order factor; 3) estimating the second-order model as conceptually and empirically warranted. We will present only the third sequence here.

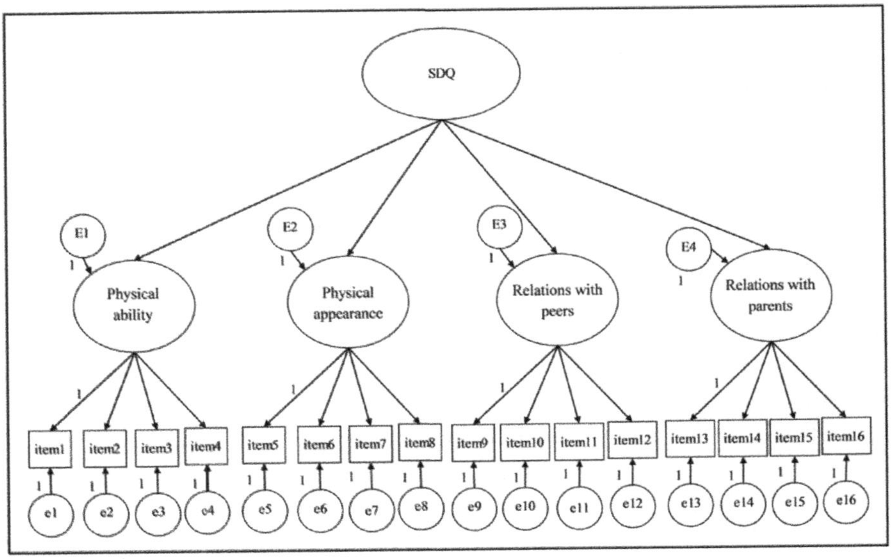

Figure 3.12. *Higher-order (second-order) CFA model The variances – curved double arrows – of the exogenous latent variables have not been represented*

3.5.1.3.1. Illustration

The following illustration has three objectives. First, to demonstrate how to use a correlation matrix accompanied by standard deviations, as the input matrix for lavaan, in the place of the unavailable raw data, to estimate a model. Next, to show how to retranscribe in lavaan syntax, estimating, and then evaluating a higher-order CFA model. Finally, comparing the hierarchical factor solution with a bifactorial solution, which will be presented later.

The correlations matrix and the standard deviations come from a paper by Marsh and Hocevar [MAR 85] in which these authors proposed to test the factorial structure of the Self-Description Questionnaire (SDQ). The SDQ is a multidimensional scale of self-concept. It measures four non-academic aspects of self-concept (physical ability, physical appearance, relations with peers, relations with parents) and three academic aspects of self-concept (reading, mathematics, general school) in children and teenagers. For the purpose of this illustration, we will restrict ourselves here to the four non-academic self-concept dimensions measured among 251 students (grade 5). Each dimension was measured by four indicators (items).

3.5.1.3.2. The correlations matrix and the standard deviations as 'input' for lavaan

Apart from the polychorical or tetrachorical correlations generated when necessary from raw data, lavaan not only analyzes the covariances matrix, but requires, above all, that these be symmetric (i.e. square matrix). However, as is often the case, in the presence of a triangle (lower or higher) of a covariances/correlations matrix, lavaan makes it possible to make this matrix symmetric. Moreover, in order to convert a correlations matrix into a covariances matrix, we must have the standard deviations of the intercorrelated variables. This conversion operation, whose technical aspects are beyond the scope of this chapter, can be easily realized using lavaan, by invoking the function "cor2cov". The details are as follows:

STEP 1. Using as imput a correlation matrix along with standard deviations of the variables.

```
# Enter or copy-paste the lower triangular matrix of correlations.

> matrix.cor <- '
1.00
.31 1.00
.52 .45 1.00
.54 .46 .70 1.00
.15 .33 .22 .21 1.00
.14 .28 .21 .13 .72 1.00
.16 .32 .35 .31 .59 .56 1.00
.23 .29 .43 .36 .55 .51 .65 1.00
.24 .13 .24 .23 .25 .24 .24 .30 1.00
.19 .26 .22 .18 .34 .37 .36 .32 .38 1.00
.16 .24 .36 .30 .33 .29 .44 .51 .47 .50 1.00
.16 .21 .35 .24 .31 .33 .41 .39 .47 .47 .55 1.00
.08 .18 .09 .12 .19 .24 .08 .21 .21 .19 .19 .20 1.00
.01 -.01 .03 .02 .10 .13 .03 .05 .26 .17 .23 .26 .33 1.00
.06 .19 .22 .22 .23 .24 .20 .26 .16 .23 .38 .24 .42 .40 1.00
.04 .17 .10 .07 .26 .24 .12 .26 .16 .22 .32 .17 .42 .42 .65 1.00
.03 .08 .08 .06 .15 .16 .10 .20 .24 .14 .21 .18 .21 .21 .29 .29 1.00
.06 .09 .11 .10 .10 .14 .11 .24 .19 .16 .23 .11 .23 .20 .32 .36 .77 1.00
.08 .15 .17 .14 .18 .20 .22 .30 .26 .24 .30 .25 .21 .25 .33 .30 .75 .74 1.00
.06 .08 .15 .18 .09 .08 .08 .15 .18 .17 .24 .17 .18 .26 .28 .24 .68 .70 .77 1.00
-.03 .30 .19 .08 .15 .09 .07 .08 .00 .13 .06 .12 .05 .10 .10 .18 .21 .09 .19 .22 1.00
.12 .33 .28 .22 .25 .14 .20 .22 .18 .33 .23 .26 .13 .18 .24 .18 .29 .19 .35 .33 .71 1.00
-.01 .37 .25 .11 .16 .09 .15 .16 .07 .22 .22 .21 .04 .08 .14 .16 .24 .20 .33 .30 .73 .79 1.00
```

```
-.01 .33 .30 .14 .17 .08 .16 .21 .09 .24 .23 .21 .09 .10 .12 .18 .24 .18 .28 .30 .75 .75 .87 1.00
.01 .19 .11 .07 .15 .17 .13 .19 .17 .24 .27 .15 .05 .11 .14 .20 .39 .37 .41 .33 .40 .48 .50 .49 1.00
.17 .35 .30 .22 .21 .23 .19 .24 .22 .28 .24 .22 .06 .04 .16 .21 .35 .35 .34 .32 .39 .47 .49 .47 .49 1.00
-.02 .16 .16 .12 .14 .12 .15 .15 .11 .18 .28 .18 .16 .23 .36 .31 .44 .45 .49 .51 .49 .58 .59 .54 .60 .39 1.00
.07 .27 .23 .19 .14 .11 .18 .21 .16 .20 .32 .21 .16 .17 .27 .31 .46 .46 .53 .46 .42 .52 .54 .52 .62 .52 .68 1.00'

# Making the matrix symmetric with the name of each of the 28 variables.

>matrix.sym <- getCov (matrix.cor, names = c ("item1",
"item2", "item3", "item4", "item5", "item6", "item7",
"item8", "item9", "item10", "item11", "item12", "item13",
"item14", "item15, "item16"," item17", "item18", "item19",
"item20", "item21", "item22", "item23", "item24", "item25",
"item26", "item27"," item28"))

# Converting the symmetric correlations and standard deviations (sds) matrix into
covariances matrices by using the function "cor2cov".

>SDQ.cov <- cor2cov (matrix.sym, sds = c (1.84, 1.94, 2.07,
1.82, 2.34, 2.61, 2.48, 2.34, 1.71, 1.93, 2.18, 1.94, 1.31,
1.57, 1.77, 1.47, 2.10, 2.11, 2.23, 2.23, 2.47, 2.21, 2.46,
2.36, 2.00, 1.84, 2.26, 2.13))

# Visualizing the covariances matrix thus obtained.

>SDQ.cov     # Clicking on enter to visualize
```

The use of a correlation matrix along with standard deviations as input involves three simple operations: 1) first of all, entering or copy-pasting the matrix (which in our example is called "matrix.cor") and placing it within single quotation marks (' '); 2) transforming the triangular matrix into a symmetric matrix and naming its constituent variables in the order that they appear within the matrix using the function "getCov"; the name of the matrix entered earlier is within parentheses and the name of each variable is entered within quotation marks "names = c ("", "", etc.)"; 3) converting the correlations matrix into a covariances matrix by using the standard deviations of the variables ("sds") and by using the function "cor2cov". To ensure that the conversion has indeed taken place, just type the name given to the converted dmatrix ("SDQ.cov ", in our example) after the chevron > and then click on enter.

It must be specified here that the conversion operation is not required if the matrix that has been entered or copy-pasted is a covariance matrix and not a correlation matrix.

Finally, it can be noted that in our example, the matrix that will be specified (in step 3) in order to be analyzed to estimate the model is called "SDQ.cov ".

3.5.1.3.3. Model specification in lavaan syntax

The two new features that must be pointed out in the specification below relate to step 3, where "data=" gives way to "sample.cov=" to indicate the name given to the matrix that has been converted and that will be used for estimating the model ("SDQ.cov" in our example) and to "sample.nob=" to specify the number of observations (sample size = 251 participants, in our example). We can also note that only 16 items of the 28 variables listed in the correlations matrix will be used in order to test the present model.

STEP 2.– Specification of the second-order SDQ model (Figure 3.12).

```
model.SPE <- 'ability =~ item1 + item2 + item3 + item4
appearance =~ item5 + item6 + item7 + item8
peers =~ item9 + item10 + item11 + item12
parents =~ item13 + item14 + item15 + item16
SDQ = ~ ability + appearance + peer + parents'
```

STEP 3. Model estimation using the model-fitting function "cfa".

```
model.EST <- cfa (model.SPE, sample.cov = SDQ.cov,
sample.nobs = 251)
```

STEP 4. Retrieving the results inlcuding the modification indices.

```
summary (model.EST, fit.measures = TRUE,
standardized = TRUE, rsq = TRUE)
```

We can again observe that for the identification of the model, the saturation of the first item of each of the four lower-order dimensions is fixed at 1.00 by default (item 1 for the latent variable "ability", item 5 for "appearance", item 9 for "peers", item 13 for "parents"). Similarly, the effect of the SDQ on the first-order latent variable "ability" is fixed at 1.00. Setting a value like this occurs by default, as "ability" figure first in the specified structural equation (SDQ = ~ ability + appearance + peers + parents).

Finally, we note that by default (and in this case, out of compulsion) the estimation method will be ML as we have a covariance matrix as input.

3.5.1.3.4. Evaluation of the solution

A review of the solution will return to the results sought in "summary", namely the model fit indices (fit.measures), the standardized estimates (standardized = TRUE) and the R^2 (rsq = TRUE).

Overall goodness-of-fit indices

Inspection of the solution shows that the overall model fit is not very satisfactory but remains acceptable (Table 3.8). In effect not only the χ^2 value was, as expected, statistically significant, but also TLI (0.909) and RMSEA (0.069) values do not plead in favor of the hierarchical solution.

```
lavaan (0.5-23.1097) converged normally after  51 iterations

  Number of observations                           251

  Estimator                                         ML
  Minimum Function Test Statistic              219.475
  Degrees of freedom                               100
  P-value (Chi-square)                           0.000

Model test baseline model:

  Minimum Function Test Statistic             1691.802
  Degrees of freedom                               120
  P-value                                        0.000

User model versus baseline model:

  Comparative Fit Index (CFI)                    0.924
  Tucker-Lewis Index (TLI)                       0.909

Loglikelihood and Information Criteria:

  Loglikelihood user model (H0)              -7584.788
  Loglikelihood unrestricted model (H1)      -7475.051

  Number of free parameters                         36
  Akaike (AIC)                               15241.576
  Bayesian (BIC)                             15368.493
  Sample-size adjusted Bayesian (BIC)        15254.368

Root Mean Square Error of Approximation:

  RMSEA                                          0.069
  90 Percent Confidence Interval          0.057  0.081
  P-value RMSEA <= 0.05                          0.007

Standardized Root Mean Square Residual:

  SRMR                                           0.060
```

Table 3.8a. *Goodness-of- fit indices of the higher-order CFA model (Figure 3.12)*

The local fit indices of the solution

First of all, the solution suffers from no anomaly (Table 3.8b). Next, two things draw our attention. First, the standardized factorial factor loadings of the indicators by the first-order factors (i.e. lower-order factor loadings), all of which go beyond 0.50.→ Secondly, the effects of the second-order factor on the first-order factors (i.e., higher-order factor loadings), namely, the four dimension of self-concept: "physical ability", "appearance", "peers" and "parents". → The effect of the SQDQ on "physical ability" is 0.545; it is 0.735 on "appearance", 0.880 on "peers" and 0.503 on "parents". All these effects are statistically significant ($p = 0.000$).

```
Parameter Estimates:

  Information                                 Expected
  Standard Errors                             Standard

Latent Variables:
                   Estimate  Std.Err  z-value  P(>|z|)   Std.lv  Std.all
  ability =~
    item1             1.000                                1.139    0.620
    item2             0.933    0.128    7.309    0.000    1.063    0.549
    item3             1.530    0.156    9.813    0.000    1.743    0.844
    item4             1.326    0.136    9.773    0.000    1.510    0.831
  appearance =~
    item5             1.000                                1.862    0.797
    item6             1.072    0.087   12.344    0.000    1.996    0.766
    item7             1.035    0.082   12.562    0.000    1.927    0.779
    item8             0.942    0.078   12.082    0.000    1.755    0.751
  peers =~
    item9             1.000                                1.023    0.599
    item10            1.214    0.155    7.831    0.000    1.242    0.645
    item11            1.668    0.190    8.790    0.000    1.706    0.784
    item12            1.364    0.162    8.413    0.000    1.395    0.720
  parents =~
    item13            1.000                                0.704    0.538
    item14            1.160    0.186    6.224    0.000    0.816    0.521
    item15            2.035    0.260    7.839    0.000    1.432    0.811
    item16            1.652    0.211    7.813    0.000    1.162    0.792
  SDQ =~
    ability           1.000                                0.545    0.545
    appearance        2.202    0.405    5.434    0.000    0.735    0.735
    peers             1.448    0.290    4.993    0.000    0.880    0.880
    parents           0.570    0.135    4.208    0.000    0.503    0.503

Variances:
                   Estimate  Std.Err  z-value  P(>|z|)   Std.lv  Std.all
   .item1             2.075    0.207   10.003    0.000    2.075    0.615
   .item2             2.619    0.252   10.384    0.000    2.619    0.699
   .item3             1.231    0.204    6.039    0.000    1.231    0.288
   .item4             1.019    0.158    6.437    0.000    1.019    0.309
   .item5             1.987    0.244    8.129    0.000    1.987    0.364
   .item6             2.802    0.322    8.706    0.000    2.802    0.413
   .item7             2.411    0.284    8.492    0.000    2.411    0.394
   .item8             2.375    0.266    8.931    0.000    2.375    0.435
   .item9             1.867    0.189    9.864    0.000    1.867    0.641
   .item10            2.168    0.228    9.492    0.000    2.168    0.584
   .item11            1.824    0.249    7.326    0.000    1.824    0.385
   .item12            1.804    0.210    8.573    0.000    1.804    0.481
   .item13            1.214    0.119   10.183    0.000    1.214    0.710
   .item14            1.789    0.174   10.275    0.000    1.789    0.729
   .item15            1.070    0.180    5.956    0.000    1.070    0.343
   .item16            0.801    0.124    6.478    0.000    0.801    0.372
    ability           0.912    0.190    4.792    0.000    0.703    0.703
    appearance        1.595    0.335    4.767    0.000    0.460    0.460
    peers             0.237    0.114    2.071    0.038    0.226    0.226
    parents           0.370    0.092    4.037    0.000    0.747    0.747
    SDQ               0.386    0.124    3.117    0.002    1.000    1.000
```

Table 3.8b. *Parameter estimates of the model in Figure 3.12 (contd.)*

Examining the variances and the R^2 (Table 3.8b and Table 3.8c) allows us to determine, for each first-order dimension, what portion of its variance can be imputed to the second-order factor (SDQ). We can observe, for example, that the variance (column "std.all") of the "physical ability" dimension is equal to 0.703. Using this estimation, we can see that the SDQ factor accounts for 30% here ((1 – 0.703 = 0.297). The value 0.297 (rounded off to 30%) corresponds to R^2. We can also see that the part of the variance of the dimension "relations with peers" that can be imputed to the SDQ hierarchical factor is 77%. For the dimension "relations with parents" this is only 25%.

```
R-Square:
                   Estimate
    item1             0.385
    item2             0.301
    item3             0.712
    item4             0.691
    item5             0.636
    item6             0.587
    item7             0.606
    item8             0.565
    item9             0.359
    item10            0.416
    item11            0.615
    item12            0.519
    item13            0.290
    item14            0.271
    item15            0.657
    item16            0.628
    ability           0.297
    appearance        0.540
    peers             0.774
    parents           0.253
```

Table 3.8c. *The R^2 of the solution (contd.)*

3.5.1.4. *The bifactorial representation of a measure*

This representation offers an interesting alternative to the hierarchical second-order representation. The two representations are conceptually similar, but functionally and mathematically different. The bifactorial representation hypothesizes that each indicator of a measure is directly dependent on three sources of influence: (1) a common factor, (2) a specific factor (dimension) for which the indicator is the unique representative, and (3) the measurement error. Figure 3.13 presents the bifactorial SDQ model. It can be noted that in the bifactorial model, the common factor is not correlated with specific factors which, in turn, are not correlated among themselves; this cannot fail to surprise, and rightly so.

Chen, West and Sousa [CHE 06] had presented the multiple benefits of the bifactorial model, including the ability to work with models that had less than four dimensions, unlike the second-order representations. However, its main advantage is that of helping judge the relevance of the presence of the sub-dimensions in a measure. Given that the factors present are orthogonal, the factorial factor loadings are instructive. In effect, if the factorial factor loadings on the common factor are high while those on the specific factors prove to be weak, then the presence of the latter is problematic. On the other hand, if the loadings on the common factor are weak, while those on the specific factors are strong, multidimensionality is seen to be justified. High loadings on the common factor as well as the specific factors argue in favor of the utility of using the total score as well as the scores of the sub-dimensions. As we have already seen, even though there is no golden rule, a factorial factor loading ≥ 0.40 is considered as being required to judge the quality of an item, to be a good indicator of the factor on which it is dependent. It must be specified that this modeling is better adapted to constructs that are assumed to be unidimensional, but with sub-dimensions that are quite substantial, which must be specified in order to reach the necessary model fit. Nonetheless, these models quite often risk encountering convergence problems.

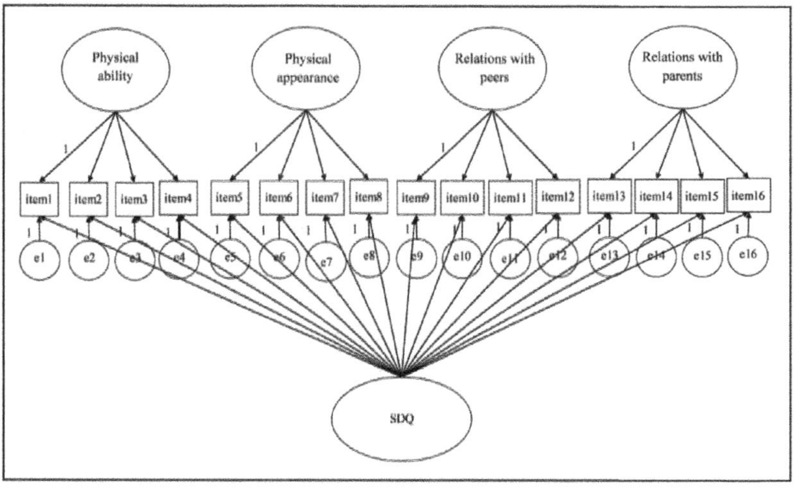

Figure 3.13. *Bifactorial measurement model (the variances – curved double arrows – of the exogenous latent variables have not been represented)*

3.5.1.4.1. Illustration

The SDQ will be used here as an example, allowing us to compare two alternative factorial solutions: hierarchical and bifactorial. In effect, the bifactorial model could be used as a basic model with which the second-order model can be compared. These two models are nested [YUN 99], in the sense that the second-order model, with fewer parameters, could be considered a special case of the bifactorial model (the second-order model is nested under the bifactorial model). Thus, the difference between the values and their χ^2 constitute a statistical test (($\Delta\chi^2 = \chi^2$ second order – χ^2 bifactorial) with the number of degrees of freedom being the difference between their respective degrees of freedom ($\Delta df = df$ second order – df bifactorial). A non-significant difference indicates that the fit of both models is similar. When, on the other hand, this difference is statistically significant, it can be inferred that the bifactorial model fits the data substantially better than the second-order model.

We must also note that it is customary to test a first-order (oblique) multi-dimensional model before proceeding to the estimation of the bifactorial model. This step will be ignored here to proceed directly to the estimation of the bifactorial model.

3.5.1.4.2. Specification of the bifactorial model in lavaan syntax

The two specificities with respect to the preceding model are: 1) each of the 16 items loads on the SDQ factor in parallel to their loading on their respective factor; 2) all the correlations between the factors (estimated by default with lavaan) are constrained to be zero and specified as such. For example, "ability~~0*appearance" signifies that the correlation between these two dimensions is constrained to be zero (non-existent).

STEP 2. Specification of the bifactorial SDQ model (Figure 3.13)

```
model.SPE <- 'ability =~ item1 + item2 + item3 + item4
appearance =~ item5 + item6 + item7 + item8
peers =~ item9 + item10 +item11 + item12
parents =~ item13 + item14 + item15 + item16
SDQ =~ item1 + item2 + item3 + item4 + item5 + item6 +
item7 + item8 +
item9 + item10 + item11 + item12 + item13 + item14 +
item15 + item16
ability ~~ 0* * appearance
ability ~~ 0*peers
ability ~ ~ 0* * parents
appearance ~ ~ 0* * peers
appearance ~~ 0*parents
peers ~ ~ 0*parents
SDQ ~~ 0*ability
```

```
SDQ ~~ 0*appearance
SDQ ~~ 0*peers
SDQ ~~ 0*parents'
```

STEP 3. Model estimation using the model-fiiting function "cfa".

```
model.EST <- cfa (model.SPE, sample.cov = SDQ.cov,
sample.nobs = 251)
```

STEP 4. Retrieving the results including the modification indices.

```
summary (model.EST, fit.measures = TRUE,
standardized = TRUE, rsq = TRUE)
```

3.5.1.4.3. Evaluation of the solution

Using the ML (*maximum likelihood*) estimator, the solution converged normally after 89 iterations (see Table 3.9). Moreover, it suffers from no offending estimates, such as a negative variance. Tables 3.9a and 3.9b provide the details of the solution.

Overall goodness-of-fit indices

If we read only the χ^2 whose value is equal to 157.59 for 88 degrees of freedom significant at $p = 0.000$, the model's fit to data cannot be accepted. Nonetheless, the other fit indices ((TLI = 0.940, CFI = 0.956, RMSEA = 0.056) argue in favor of the bifactorial solution.

It will now be instructive to compare the fit of the bifactorial solution with the fit of the second-order solution (see Table 3.8). The difference in the values of their χ^2 is 61.88 (= 219.47 – 157.59) with a difference of 12 *df* (100-88). This difference ($\Delta\chi^2(12)$ = 61.88) is significant at p$<$*0.05 (it is enough to consult a* χ^2 table), indicating that the bifactorial solution best approximates the reality of than the hierarchical solution.

Local fit indices

Table 3.9 gives the details. The review will focus on the comparison of the factor loadings, either on the genetal factor (SDQ) or on the specific factors ("ability", "appearance", "peers", "parents"). The logic is simple. If the factor loadings on specific factors are higher and more substantial than those on the general factor, the multidimensionality of the measure is reinforced. Conversely, the unidimensionality is reinforced.

```
lavaan (0.5-23.1097) converged normally after  89 iterations

  Number of observations                           251

  Estimator                                         ML
  Minimum Function Test Statistic              157.659
  Degrees of freedom                                88
  P-value (Chi-square)                           0.000

Model test baseline model:

  Minimum Function Test Statistic             1691.802
  Degrees of freedom                               120
  P-value                                        0.000

User model versus baseline model:

  Comparative Fit Index (CFI)                    0.956
  Tucker-Lewis Index (TLI)                       0.940

Loglikelihood and Information Criteria:

  Loglikelihood user model (H0)              -7553.880
  Loglikelihood unrestricted model (H1)      -7475.051

  Number of free parameters                         48
  Akaike (AIC)                               15203.760
  Bayesian (BIC)                             15372.982
  Sample-size adjusted Bayesian (BIC)        15220.816

Root Mean Square Error of Approximation:

  RMSEA                                          0.056
  90 Percent Confidence Interval          0.042  0.070
  P-value RMSEA <= 0.05                          0.228

Standardized Root Mean Square Residual:

  SRMR                                           0.055
```

Table 3.9a. *Goodness-of-fit indices for the bifactorial SDQ model (Figure 3.13)*

We can see that of the 16 factor loadings by the specific factors, 13 show factor loadings greater than 0.40. Only 10 saturations by the general factor display values equal to or greater than 0.40. Moreover, the mean saturation (computed manually) by the specific factors comes to 0.527, while the average factor loading on the general factor is 0.458. The multidimensionality of the scale seems more likely than the existence of a general factor underpinning (or explaining, in the second-order model) an underlying structure.

```
Parameter Estimates:

  Information                                 Expected
  Standard Errors                             Standard

Latent Variables:
                   Estimate  Std.Err  z-value  P(>|z|)   Std.lv  Std.all
  ability =~
    item1             1.000                               1.073    0.584
    item2             0.664    0.133    4.999    0.000    0.713    0.368
    item3             1.227    0.156    7.864    0.000    1.316    0.637
    item4             1.221    0.157    7.767    0.000    1.310    0.721
  appearance =~
    item5             1.000                               1.580    0.677
    item6             1.081    0.219    4.930    0.000    1.709    0.656
    item7             0.487    0.110    4.434    0.000    0.769    0.311
    item8             0.283    0.104    2.714    0.007    0.447    0.191
  peers =~
    item9             1.000                               0.885    0.519
    item10            0.946    0.224    4.231    0.000    0.838    0.435
    item11            1.023    0.229    4.474    0.000    0.906    0.416
    item12            1.077    0.238    4.517    0.000    0.953    0.492
  parents =~
    item13            1.000                               0.621    0.475
    item14            1.291    0.235    5.504    0.000    0.802    0.512
    item15            1.987    0.303    6.559    0.000    1.234    0.699
    item16            1.763    0.270    6.523    0.000    1.095    0.746
  SDQ =~
    item1             1.000                               0.487    0.265
    item2             1.631    0.439    3.714    0.000    0.795    0.411
    item3             2.264    0.524    4.322    0.000    1.104    0.534
    item4             1.687    0.383    4.402    0.000    0.822    0.453
    item5             2.596    0.763    3.402    0.001    1.265    0.542
    item6             2.697    0.808    3.339    0.001    1.315    0.505
    item7             3.608    0.989    3.648    0.000    1.758    0.710
    item8             3.743    1.017    3.681    0.000    1.825    0.781
    item9             1.324    0.420    3.153    0.002    0.645    0.378
    item10            1.834    0.544    3.370    0.001    0.894    0.464
    item11            2.949    0.811    3.636    0.000    1.437    0.661
    item12            2.146    0.613    3.500    0.000    1.046    0.540
    item13            0.654    0.252    2.597    0.009    0.319    0.244
    item14            0.447    0.255    1.755    0.079    0.218    0.139
    item15            1.412    0.439    3.217    0.001    0.688    0.390
    item16            0.947    0.320    2.957    0.003    0.462    0.315
```

Table 3.9b. *Local indices (parameter estimates) of the bifactorial SDQ model (Figure 3.13) (contd.)*

Finally, the R^2 ("R-square") values for each item vary from 0.281 (for item 14) to 0.751 (for item 5). Where item 14 is concerned, 28% of its variance can be imputed as much to the general factor (SDQ) as to the specific factor on which it depends ("parents"). For item 5, 75% of its variance can be attributed either to the general factor (SDQ) or to the specific factor on which it depends ("appearance").

```
R-Square:
                   Estimate
    item1             0.412
    item2             0.304
    item3             0.691
    item4             0.725
    item5             0.751
    item6             0.685
    item7             0.601
    item8             0.647
    item9             0.412
    item10            0.404
    item11            0.610
    item12            0.534
    item13            0.285
    item14            0.281
    item15            0.640
    item16            0.656
```

Table 3.9c. *Local indices of the bifactorial SDQ model (here the R^2). (Figure 3.13) (contd.)*

Reflective measurement model versus formative measurement model

In conclusion, it seems useful to raise the question of the status of indicators within measurement models: are they the effect (consequence) or the cause of the latent variable? Until now, we have presented them as being subject to the influence of latent variables (reflective model).

In effect, it is assumed that this influence explains the interdependence that exists between these indicators. This conception is reflected in a graph where the arrows move from factors towards the measured variables (items). The monofactorial representation of the life satisfaction scale is an illustration of this. As reflective indicators, these items reflect the effect of the factor (latent factor) on which they depend. Thus, each reflective item is accompanied by a measurement error which is assumed to be non-correlated with the latent factor. However, according to Bollen and Lennox [BOL 91], there is nothing to prevent us from considering the indicators as being the source of a latent variable (also see [EDW 00]). Indeed, it is not absurd to think that the indicators could explain a latent factor. For example, we could assume that satisfaction with life depends, among other things, on subjective health, financial conditions, marital satisfaction, satisfaction at work and leisure activities. Figure 3.14 shows this point of view with respect to the determinant (causal) indicators. This point of view seems, in many cases, likelier than thinking of determined indicators (effects). This is a formative measurement model.

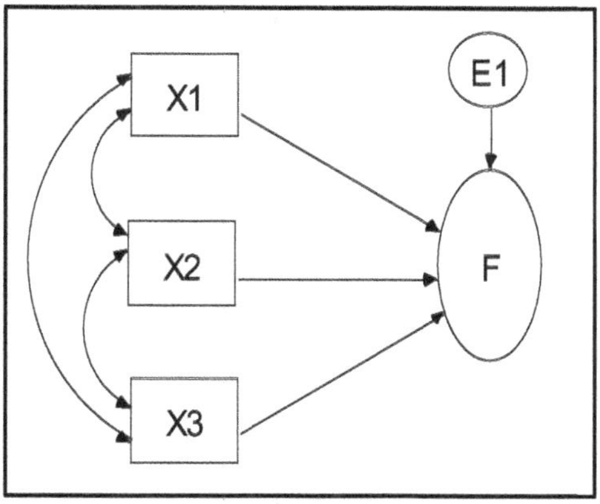

Figure 3.14. *Formative measurement model with causal indicators*

Here, it must be emphasized that formative measurement models quite often encounter identification and convergence problems [DIA 08, GRA 08]. Difficulties such as these may be resolved by integrating these models within structural models made up of reflective latent variables. In effect, the models multiple indicators and multiple causes (MIMIC), put forth by Jöreskog and Sörbom [JÖR 96], offer a heuristic compromise that may resolve the identification and convergence problems. In its simplest form, a MIMIC model contains a single latent variable (factor) that is dependent on and determined by certain measured variables (the causal indicators) on the one hand, while explaining a number of observed variables (effects-indicators) on the other hand. Figure 3.15 graphically illustrates such a model. The reader can find in Loehlin [LOE 98] a more complex illustrated form that brings in several latent variables.

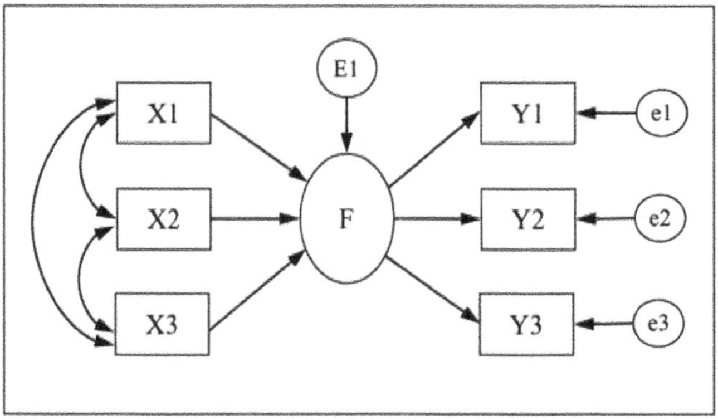

Figure 3.15. *The multiple indicators and multiple causes (MIMIC) model*

It is very simple to retranscribe a formative or MIMIC measurement model into lavaan syntax:

STEP 2. Specification of the formative measurement model (Figure 3.14).

```
model.SPE <- 'F <~ x1 + x2 + x3'
```

STEP 2. Specification of the MIMIC model (Figure 3.15).

```
model.SPE <- 'F =~ Y1 + Y2 + Y3
              F <~ x1 + x2 + x3'
```

Or again, by using the regression operator ~ instead of the formative operator <~.

```
model.SPE <- 'F =~ Y1 + Y2 + Y3
              F ~ x1 + x2 + x3'
```

3.5.1.5. *Structural Model*

As Figure 3.8b shows, the structural model refers to that portion of the general model that contains the relationships between latent variables. The structural model, strictly speaking, refers to the interrelation between the latent variables within a general model containing latent variables. This is actually the part relating to the regression between these latent variables. Thus, each endogenous variable is expressed as a linear function of the variables that predict it. There are, therefore, as many structural equations as there are endogenous (dependent) variables in a model. Let us return to Figure 3.8b, which we will use to illustrate these observations. Figure 3.16 is an extraction of the structural portion. This portion of a general structural model contains four latent variables: two exogenous (independent) correlated variables ("self-esteem" and "optimism") and two endogenous (dependent) variables ("self-stereotype" and "health") that give rise to two structural equations.

$$F3 = F1 + F2 + E1$$

$$F4 = F3 + E2$$

As we have seen, each equation contains as many terms as there are arrows pointing to the concerned dependent variable. To put it differently, we could say, for example, that F3 is subject to the direct predictive effects of F1 and F2 as well as the influence of an error term ("E1"), the variable that represents the sources of variations (disturbances) external to the model.

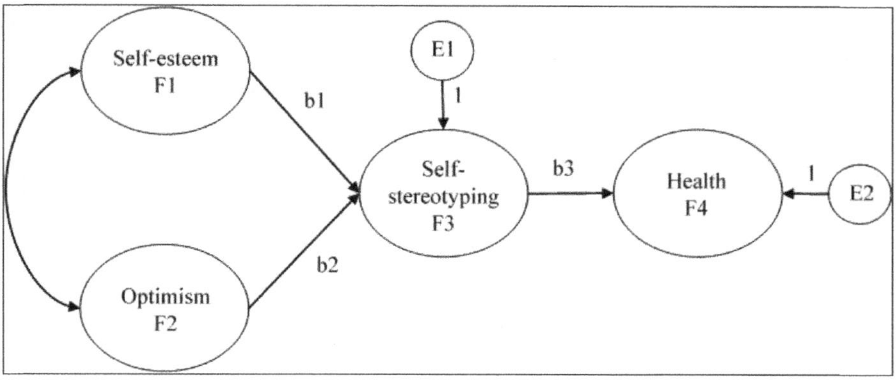

Figure 3.16. *Structural portion of Figure 3.8b*

Each term in the equation represents a direct predictive effect on the dependent variable in question. What about the indirect effects?

Here also, of course, we have the total effect of a latent variable on another latent variable being equal to the sum of the direct effect and/or all indirect effects of one on the other, this generally being the product of the other direct effects. Here is an illustration from the graph in Figure 3.16:

F3 → F4 = direct effect

F1 ---> F4 = indirect effect of F1 on F4 via F3

F2 ---> F4 = indirect effect of F2 on F4 via F3

The indirect effect of F1 on F4 via F3 is obtained by multiplying the path coefficients b1 and b3 (b1*b3); and the indirect effect of F2 on F4 via F3 is the product of the path coefficients b2 and b4 (b2*b3). Thus, the total effect on F4 is equal to (b1*b3) + (b2*b3) + b3.

As we have seen, contrary to most modeling software, lavaan does not automatically compute these different effects. However, the mathematical aspects of these computations will not be described here. For this, we refer the reader to Mueller's text [MUE 96].

Moreover, recursiveness or non-recursiveness is a property that also concerns structural models with latent variables. Three types of relationship, illustrated in

Figure 3.17, generate the non-recursiveness of a model: the causal reciprocity between at least two variables, the feedback loop between several variables and the correlation between residual variables within a model [RIG 95].

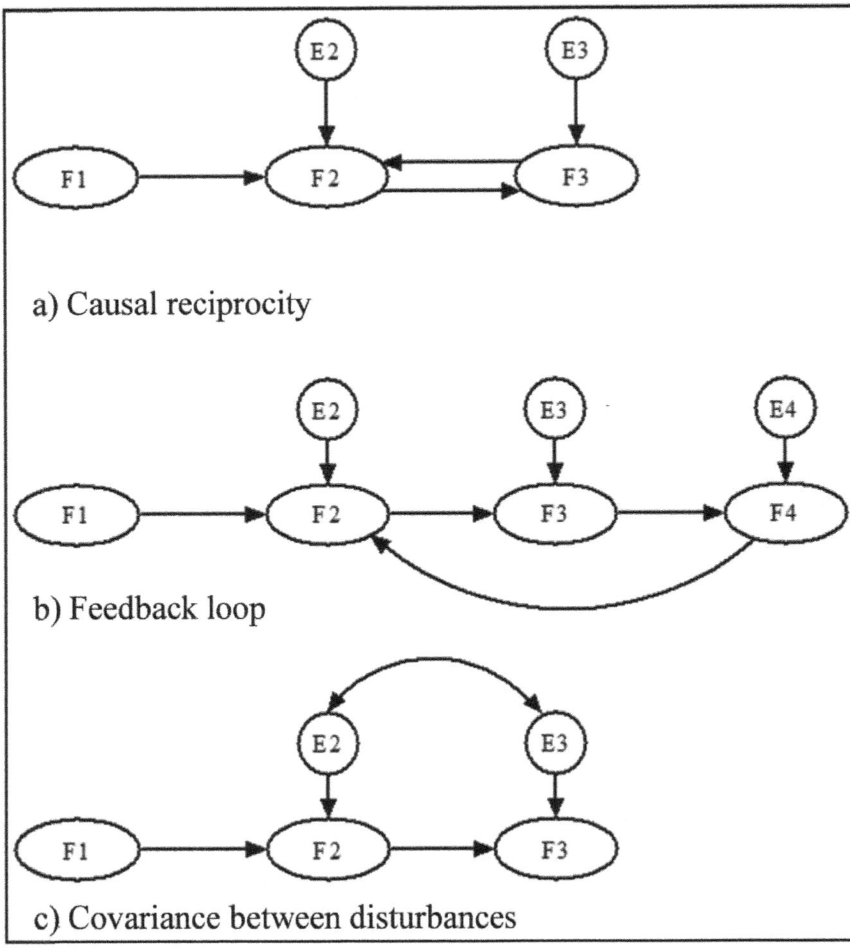

Figure 3.17. *Three types of relationship generating non-recursiveness*

3.5.1.5.1. Identification of a structural model with latent variables

The identification of a structural model is often complex because of the coexistence of the latent and the measured variables, as this concerns the general model and not only the portion relating to the relationship between latent variables. The main difficulty is determining and counting the parameters that are free to be estimated within a model.

To obtain the number of the degrees of freedom of the model, we simply need to substract the number of parameters to be estimated from the number of variances and covariances of the measured variables.

3.5.1.5.2. Model estimation

As soon as it is over-identified, a model can be testable. The question then arises: which approach to estimation we have to choose: a one-step modeling approach or a two-step modeling approach?

Model estimation in one or two steps?

The question of how many steps the estimation procedure should comprise comes up only in the presence of a general structural model, which, we must recall, is a hybrid combination of measurement and structural models. A simple example can help clarify this concept. Let us assume that we have a general structural model that fails to fit our data. A legitimate and worrying question then arises about the reasons for such a failure: is this imputable to the measurement models that constitute it – and to them alone?; is it imputable to the structural model – and to this alone?; or, perhaps, is it imputable to the entirety of this model? It is clear that one single estimation will not allow us to answer this. Thus, to reach an answer, we must proceed step by step, that is, carry out two separate analyses: the first will focus on the measurement model and the second on the specified general structural model, and this will happen only if the first model proves to be satisfactory. The first analysis is thus considered as a prerequisite for the estimation of the general structural model.

Initially proposed by James *et al.* [JAM 82] and then presented in a clearly argued form by Anderson and Gerbing [AND 88], this procedure is now of working is now known as the two-step approach. However, despite the apparent common sense that seems to characterize this, it has not been unanimously accepted by specialists. An animated debate was sparked off between specialists such as Anderson and Gerbin [AND 88] and Mulaik and James [MUL 95] (the major arguments in this debate can be found in Hayduk [HAY 96]) who believed that this procedure was not only justified but necessary, and other specialists such as Fornell and Yi [FOR 92a, FOR 92b] and Hayduk [HAY 96], who refused to admit that this procedure was in any way useful. This last author claimed that this debate was a temporary distaction that will dissipate once our attention focuses on more useful subjects [HAY 96].

Let us first take a good look at what constitutes the two-step approach before we examine why it is criticized. The basic idea of this approach is based on the consideration that it is necessary to ensure the validity of the measures before testing the models that use them. A model is, largely, only as good as its measures. Thus, Anderson and Gerbing [AND 88] suggested that we consider that the measurement model provides an assessment of the convergent and discriminant validity of the measurement tools used while the structural model provides the predictive validity. In other words, the first tests the hypotheses related to the structures of the measured constructs that come from robust theoretical definitions of these constructs, while the second assesses hypotheses on the relationships between these constructs, arising from the nomological network theories about these constructs. The logic is then quite simple (perhaps even too simple): moving to the nomological perspective (validity) imperatively requires that the theoretical definitions of the constructs be valid (convergent validity, divergent validity, reliability), a condition that is assumed as the prerequisite. Let us take the model presented in Figure 3.7 and assume that we wish test it. This is, clearly, a general structural model. The question that then arises is: which measurement model do we estimate in the first place? It already contains three, presented in Figure 3.8. Must we then evaluate them separately or more simply, do we group together into a single factorial model, where all the factors are allowed to covary with each other? Figure 3.18 graphically illustrates this option. The first solution is not free of difficulties, especially that of being unable to estimate (because of their under-identification) the increasingly frequent constructs containing three indicators, sometime only two (Figure 3.8c) or even a single indicator. Jöreskog and Sörbom [JÖR 93] went even further: they recommended that in the framework of this first step we first evaluate all the constructs separately, then all the possible pairs of constructs, and finally all the constructs that are interrelated with no constraint (Figure 3.18). However, in practice, we often limit ourselves to this last option, as it makes it possible to provide an instructive comparison between the measurement model and the initial structural model, a model considered to be nested within the first. As long as the structural portion is not just-identified (thus making both models equivalent and making it impossible to compare them) the difference between their respective χ^2 is highly informative. For instance, a non-significant difference when the measurement model is seen to be unsatisfactory would show that the structural model does not aggravate the model misfit. This signifies that the structural model may not be challenged, but this is not certain as, according to this approach, its success remains closely tied with the validity of the measures that it uses.

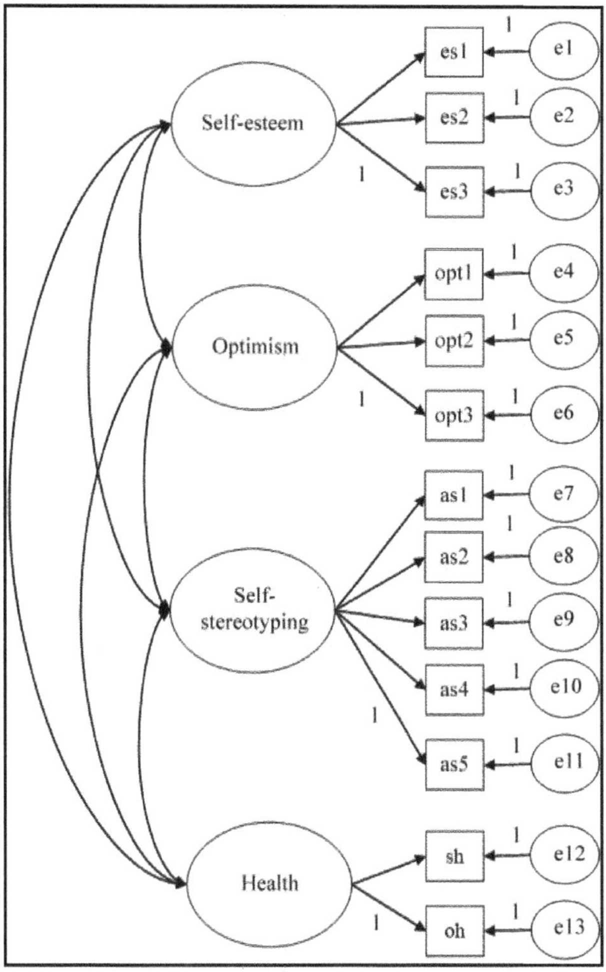

Figure 3.18. *Respecification of Model 3.7 using CFA to serve as a measurement model in the 1st step in the estimation of the general structural model (the variances – curved double arrows – of the exogenous latent variables have not been represented)*

What remains to be seen now is what must be done when the measurement model is judged to be unsatisfactory. Simply abandining the initial model is discarded, thus we have the possibility to improve it. While certain modifications allow this, none of them is miraculous. On the contrary, however, they must necessarily be accompanied by solid theoretical considerations. These revisions could affect both the indicators as well as the factors. At the level of the indicators, it is recommended that those whose loading on a factor is low be deleted (i.e. those which do not seem suitable to the construct) without, however, neglecting the possibility that these indicators could also cross-load on multiple factors. In effect, it is not uncommon for two factors to share the same indicator. Another solution that is often chosen to improve a measurement model consists of allowing correlation between the indicator error variances (i.e., correlated errors). We state here that two or more indicators may share the same unknown sources of variation. For example, their redundancy may be one of the possible causes [RUB 95]. As regards factors, their numbers are most often affected by the revisions. We recommend, of course, that this be scaled down when some strong covariations arise between the factors as this is considered to be symptomatic of rather poor discriminant validity, that is, a sign that certain constructs overlap with others. On the other hand, this number may be scaled up when several indicators are weakly loaded on the factors with which they are related. This is symptomatic, here, of convergent validity problem that affects the content of measured constructs. This means that all the sources of variation in the responses to the indicators are not representesd in the measurement model, hence the need to add one or more factors.

All these revisions, with the exception of the addition or removal of latent factors, are greatly facilitated by turning to modification indices, which are available in most modeling software. These suggest, as their name indicates, all modifications that could improve the overall fit of the model in question. The EQS software ([BEN 95] offers two important tests: the Lagrange Multiplier Test, which suggests adding parameters that contribute significantly to the model fit, and the Wald Test, which suggest the removal of parameters that do not contribute to this fit. However, if this is not achieved, it would be better to then simply abandon the initial model. The second stage of the estimation will then not take place.

Let us now turn to the critics. The legitimacy of the modifications in a confirmatory procedure is brought up from the outset. The few *post hoc* theoretical justifications offered for modifications cannot give the border between the exploratory process and the confirmatory process the illumination that it deserves. Instead, certain purists recommend the use of the duly validated measuring instruments. They express surprise, moreover, at the fact that we can dissociate measure and theory as the two-step approach implies. This is another fundamental criticism. But this is not a new problem and it remains very complex. In effect, if the measure is a fundamentally theoretical process [DIC 94], it is legitimate to wonder about the appropriateness of a distinction between theoretical definitions of constructs and nomological network of these constructs, both of which underpin this measure, especially when, as in this case, we place ourselves in a confirmatory perspective. Moreover, as Dickes *et al.* [DIC 94] indicate, these two theories do not exclude each other and they can, above all, be estimated simultaneously through SEM.

Illustration

If the available data and the sample size allow, the researcher could convert the path model (Figure 3.5) into a structural model with latent variables. It will thus take the form given in Figure 3.7 in which we can see that each latent variable replacing the observed variables in the model in Figure 3.5 is measured by a certain number of indicators (items). These, as well as the endogenous latent variables, are accompanied by residual variables ("e" = measurement error, "E" = disturbance. Figure 3.7 represents what can be called a standard general structural model. One of the major benefits of this model is that its estimates take into account measurement errors. The estimations obtained are thus purified as they are rid of any measurement error.

A brief presentation of the theoretical model

Let us recall that this is a model that brings into play the relationships between psychological resources (self-esteem, dispositional optimism), aging self-stereotypes and physical health among the elderly ($N = 331$). This is a model that fits within the conceptual framework of positive psychology and the sociocognitive approach of stereotypes.

Figure 3.7 presents this model that translates the hypotheses according to which a person's psychological resources influence their aging self-stereotypes. These self-stereotypes in turn exert an influence on one's physical health.

We can note that the model uses four latent variables (replacing the four observed variables in the model in Figure 3.5) and 13 observed variables. Measured by three indicators (here there are item parcels) each, both exogenous latent variables, in covariation, represent the psychological resources, namely, self-esteem and dispositional optimism. Each resource influences the latent variable representing the negative aging self-stereotypes. This endogenous latent variable, measured through five items, is considered to be a mediator variable between psychological resources and physical health (considered to be the ultimate endogenous variable) and measured, here, with two indicators, namely: objective health and subjective (perceived) health, where high scores are indicative of poor health.

The two-step approach

We have adopted the two-step approach to model estimation.

– *The measurement model*: the objective of the first step is to estimate the measurement model through a CFA that tests a factorial model that freely interrelates the four latent variables. Figure 3.18 offers an illustration and box for this following the specification in lavaan syntax.

STEP 2. Specification of the measurement model (step 1 of the two-step approach, Figure 3.18).

```
model.SPE <- 'esteem =~ es1 + es2 + es3
optimism =~ lot1 + lot2 + lot3
stereotype =~ as1 + as2 + as3 + as4 + as5
health =~ sh + oh
```

STEP 3. Model estimation using the model-fitting function "cfa".

```
model.EST <- cfa (model.SPE, data = BASE)
```

STEP 4. Retrieving the results including the modification indices.

```
summary (model.EST, fit.measures = TRUE,
standardized = TRUE, modindices = TRUE)
```

We will specify here that the estimation method used is the maximum likelihood (ML) method and this is because the multivariate normality of the data does not suffer from any serious violation, as evidenced by the Mardia coefficient whose value of 2.36 is lower than 5.00, the threshold value beyond which multivariate normality fails. We can also note that lavaan, by default, fixes the first indicator of each latent variable at 1.00 in order to identify the model.

Evaluation of the measurement model from step 1

The values of the overall fit indices of the specified measurement model are presented in Table 3.10. First of all, the solution converged normally after 70 iterations. Next, while, in order to assess the fit of the model, we stopped with χ^2, whose value here is equal to 123.38 with 59 degrees of freedom for $N = 331$, it is clear that this does not argue in favor of our measurement model. Nonetheless, it should be recalled that the inflation in the value of χ^2 could be imputed to the fact that this measure is very sensitive to sample size. Thus, it is better to look to other fit indices. In effect, an examination of these indices (for example CFI, RMSEA, SRMR) shows that this measurement model seems to fit the data well.

However, such results cannot do away with the need to examine local indices. Indeed, on consulting Table 3.10b, we note that the solution obtained suffers from no impurity such as inadmissible values or standard errors that are too large. We also note that the non-standardized estimate of the first indicator of each latent variable (that is: as1, sh, es1, lot1) is equal to 1.00 (as it was fixed, *a priori*). We can also see that all the standardized factorial saturations are statistically significant. The same holds true for correlations ("covariances") between the latent variables.

```
lavaan (0.5-23.1097) converged normally after  70 iterations

  Number of observations                              331

  Estimator                                            ML
  Minimum Function Test Statistic                 123.384
  Degrees of freedom                                   59
  P-value (Chi-square)                              0.000

Model test baseline model:

  Minimum Function Test Statistic                1559.972
  Degrees of freedom                                   78
  P-value                                           0.000

User model versus baseline model:

  Comparative Fit Index (CFI)                       0.957
  Tucker-Lewis Index (TLI)                          0.943

Loglikelihood and Information Criteria:

  Loglikelihood user model (H0)                 -5414.522
  Loglikelihood unrestricted model (H1)         -5352.831

  Number of free parameters                            32
  Akaike (AIC)                                  10893.045
  Bayesian (BIC)                                11014.713
  Sample-size adjusted Bayesian (BIC)           10913.208

Root Mean Square Error of Approximation:

  RMSEA                                             0.057
  90 Percent Confidence Interval          0.043  0.072
  P-value RMSEA <= 0.05                             0.186

Standardized Root Mean Square Residual:

  SRMR                                              0.046
```

Table 3.10a. *Overall goodness-of-fit indices of the model in Figure 3.18.*

```
Parameter Estimates:

  Information                                 Expected
  Standard Errors                             Standard

Latent Variables:
                   Estimate  Std.Err  z-value  P(>|z|)   Std.lv  Std.all
  stereotype =~
    as1               1.000                               0.184    0.442
    as2               1.886    0.272    6.942    0.000    0.347    0.695
    as3               1.485    0.236    6.302    0.000    0.274    0.551
    as4               1.929    0.276    6.993    0.000    0.355    0.711
    as5               0.999    0.176    5.688    0.000    0.184    0.456
  health =~
    sh                1.000                               0.620    0.805
    oh                1.851    0.246    7.519    0.000    1.148    0.599
  esteem =~
    es1               1.000                               1.609    0.830
    es2               1.085    0.055   19.785    0.000    1.746    0.898
    es3               1.639    0.083   19.717    0.000    2.637    0.895
  optimism =~
    lot1              1.000                               1.018    0.723
    lot2              0.758    0.100    7.569    0.000    0.771    0.535
    lot3              0.998    0.117    8.536    0.000    1.016    0.650

Covariances:
                   Estimate  Std.Err  z-value  P(>|z|)   Std.lv  Std.all
  stereotype ~~
    health            0.079    0.014    5.694    0.000    0.691    0.691
    esteem           -0.149    0.029   -5.218    0.000   -0.503   -0.503
    optimism         -0.121    0.023   -5.374    0.000   -0.648   -0.648
  health ~~
    esteem           -0.446    0.076   -5.897    0.000   -0.447   -0.447
    optimism         -0.280    0.055   -5.105    0.000   -0.443   -0.443
  esteem ~~
    optimism          0.771    0.131    5.899    0.000    0.471    0.471

Variances:
                   Estimate  Std.Err  z-value  P(>|z|)   Std.lv  Std.all
   .as1               0.140    0.012   12.067    0.000    0.140    0.804
   .as2               0.129    0.013    9.713    0.000    0.129    0.517
   .as3               0.172    0.015   11.422    0.000    0.172    0.696
   .as4               0.124    0.013    9.414    0.000    0.124    0.495
   .as5               0.129    0.011   12.005    0.000    0.129    0.792
   .sh                0.209    0.048    4.377    0.000    0.209    0.352
   .oh                2.353    0.239    9.851    0.000    2.353    0.641
   .es1               1.167    0.116   10.029    0.000    1.167    0.311
   .es2               0.729    0.101    7.240    0.000    0.729    0.193
   .es3               1.724    0.233    7.416    0.000    1.724    0.199
   .lot1              0.944    0.125    7.548    0.000    0.944    0.477
   .lot2              1.484    0.135   10.968    0.000    1.484    0.714
   .lot3              1.413    0.152    9.282    0.000    1.413    0.578
    stereotype        0.034    0.009    3.787    0.000    1.000    1.000
    health            0.385    0.062    6.181    0.000    1.000    1.000
    esteem            2.589    0.287    9.023    0.000    1.000    1.000
    optimism          1.035    0.169    6.130    0.000    1.000    1.000
```

Table 3.10b. *Local fit indices (parameter estimates) of the model in figure 3.18 (contd.)*

And although the solution seems perfectly acceptable, even allowing us to proceed to the second step in estimation, we think it would be highly useful to examine the modification indices, for didactic purposes. Reading Table 3.10, which presents a sample, shows us that freeing the correlated error between indicators "as1" and "as2" would lead to a drop in the value of the χ^2 equivalent to 13.745. A certain redundancy of indicators could be the reason for this. However, we did not judge it useful or justified to modify our initial model.

76	as1	~~	as2	13.745	0.034	0.034	0.162	0.162
77	as1	~~	as3	0.170	-0.004	-0.004	-0.019	-0.019
78	as1	~~	as4	0.881	-0.009	-0.009	-0.041	-0.041
79	as1	~~	as5	0.203	-0.004	-0.004	-0.021	-0.021
80	as1	~~	sh	2.933	-0.023	-0.023	-0.071	-0.071
81	as1	~~	oh	0.192	-0.015	-0.015	-0.019	-0.019
82	as1	~~	es1	1.099	0.026	0.026	0.033	0.033
83	as1	~~	es2	0.004	-0.001	-0.001	-0.002	-0.002
84	as1	~~	es3	0.449	-0.023	-0.023	-0.019	-0.019
85	as1	~~	lot1	2.388	0.038	0.038	0.065	0.065
86	as1	~~	lot2	1.872	-0.037	-0.037	-0.062	-0.062
87	as1	~~	lot3	0.020	-0.004	-0.004	-0.006	-0.006
88	as2	~~	as3	2.998	-0.019	-0.019	-0.075	-0.075
89	as2	~~	as4	0.071	0.003	0.003	0.012	0.012

Table 3.10c. *Modification indices for the model in Figure 3.18 (contd.)*

– *The structural model*: the above results allow us to test the structural model specified a priori in order to check if it can be tolerated by the data. The specification of this model in lavaan syntax appears in the box below.

STEP 2. Specification of the general structural model (Figure 3.7).

```
model.SPE <- '

# Measurement models

esteem =~ es1 + es2 +es3
optimism = ~ lot1 + lot2 + lot3
stereotype =~ as1 + as2 + as3 + as4 + as5
health =~ sh + oh

# Structural model

stereotype ~ esteem + optimism
health ~ stereotype'
```

STEP 3. Model estimation using the "sem" model-fitting function.

```
model.EST <- sem (model.SPE, data = BASE)
```

STEP 4. Retrieving the results including the modification indices.

```
summary (model.EST, fit.measures = TRUE,
standardized = TRUE, modindices = TRUE)
```

We can note that this specification encloses the specification of the measurement model (similar to step 1) as well as that of the structural portion of the model, which reflects three causal paths: the two predictive effects of self-esteem and optimism on aging self-stereotypes (stereotype ~ self-esteem + optimism) and the predictive effect of the latter on physical health (health ~ stereotype). To specify the measure models, the 'equal to' sign followed by a tilde (=~) is the operator of choice, while the tilde is the structural portion of the model.

We can note, and this is not innocuous, that unlike in the earlier step, the model-fitting function is no longer "cfa" but is now "sem" (*structural equation modeling*). As concerns the estimation method, the method used is always the maximum likelihood (ML) method, even if this does not appear in the specification. This is because this method is used by default.

Evaluation of the solution of the general structural model

According to the value of χ^2 available in Table 3.11, our hypothetical model is not able to adequately reproduce the observed variances-covariances matrix (χ^2 (61, N= 331) = 126.82, p = 0.000). However, we know that the χ^2 value is sensitive to sample size. Indeed, the values of the overall fit indices seem rather to argue in favor of the fact that this model approximates the data reasonably well (CFI = 0.956, RMSA = 0.057, SRMR = 0.047). Let us recall that the harmony of the model with the data is better when the last two indices are close to zero.

By converging normally after 61 iterations, a proper solution was obtained. Indeed, an examination of the local fit indices reveals no inadmissible or offending estimates value, thus allowing the analytical evaluation of the solution.

However, the most interesting result concerns the structural portion of the model. Upon examining Table 3.11 we can note that the path coefficients ("regressions") that express the influence of the variable "self-esteem" and of the variable "dispositional optimism" on "aging self-stereotypes" are statistically significant. The effect of "self-esteem" (β = – 0.283) is smaller than the effect of optimism (β = – 0.515). However, both of these go in the same, expected direction: both these positive psychological resources are negatively related to negative aging self-stereotypes. The more one displays high self-esteem and high optimism, the less likely they are to hold negative aging self-stereotypes. With regard to the predictive effect of the latter on physical health (where a high score indicates poor health) it has been shown to be statistically significant and high (β = 0.703). This is translated by the existence of a positive effect of negative aging self-perceptions on poor physical health. In other words, the more we hold negative self-perceptions with respect to our own aging, the more we are/feel in poor health. Let us not forget here the inherent limits to any cross-sectional research.

```
lavaan (0.5-23.1097) converged normally after  61 iterations

  Number of observations                           331

  Estimator                                         ML
  Minimum Function Test Statistic              126.823
  Degrees of freedom                                61
  P-value (Chi-square)                           0.000

Model test baseline model:

  Minimum Function Test Statistic             1559.972
  Degrees of freedom                                78
  P-value                                        0.000

User model versus baseline model:

  Comparative Fit Index (CFI)                    0.956
  Tucker-Lewis Index (TLI)                       0.943

Loglikelihood and Information Criteria:

  Loglikelihood user model (H0)              -5416.242
  Loglikelihood unrestricted model (H1)      -5352.831

  Number of free parameters                         30
  Akaike (AIC)                               10892.485
  Bayesian (BIC)                             11006.548
  Sample-size adjusted Bayesian (BIC)        10911.387

Root Mean Square Error of Approximation:

  RMSEA                                          0.057
  90 Percent Confidence Interval          0.043  0.071
  P-value RMSEA <= 0.05                          0.193

Standardized Root Mean Square Residual:

  SRMR                                           0.047
```

Table 3.11a. *Overall goodness-of-fit indices of the general structural model (Figure 3.7)*

```
Parameter Estimates:

  Information                                 Expected
  Standard Errors                             Standard

Latent Variables:
                   Estimate  Std.Err  z-value  P(>|z|)   Std.lv  Std.all
  stereotype =~
    as1               1.000                               0.182    0.438
    as2               1.892    0.275    6.881    0.000    0.345    0.690
    as3               1.503    0.240    6.273    0.000    0.274    0.552
    as4               1.939    0.280    6.936    0.000    0.354    0.707
    as5               1.011    0.178    5.672    0.000    0.184    0.456
  health =~
    sh                1.000                               0.626    0.812
    oh                1.819    0.247    7.357    0.000    1.138    0.594
  esteem =~
    es1               1.000                               1.610    0.831
    es2               1.085    0.055   19.801    0.000    1.746    0.899
    es3               1.637    0.083   19.719    0.000    2.636    0.895
  optimism =~
    lot1              1.000                               1.015    0.721
    lot2              0.763    0.101    7.583    0.000    0.774    0.537
    lot3              1.001    0.117    8.528    0.000    1.016    0.650

Regressions:
                   Estimate  Std.Err  z-value  P(>|z|)   Std.lv  Std.all
  stereotype ~
    esteem           -0.032    0.009   -3.652    0.000   -0.283   -0.283
    optimism         -0.093    0.019   -4.783    0.000   -0.515   -0.515
  health ~
    stereotype        2.412    0.379    6.368    0.000    0.703    0.703

Covariances:
                   Estimate  Std.Err  z-value  P(>|z|)   Std.lv  Std.all
  esteem ~~
    optimism          0.769    0.130    5.890    0.000    0.470    0.470

Variances:
                   Estimate  Std.Err  z-value  P(>|z|)   Std.lv  Std.all
   .as1               0.140    0.012   12.106    0.000    0.140    0.809
   .as2               0.131    0.013    9.904    0.000    0.131    0.524
   .as3               0.171    0.015   11.453    0.000    0.171    0.696
   .as4               0.125    0.013    9.604    0.000    0.125    0.500
   .as5               0.129    0.011   12.021    0.000    0.129    0.792
   .sh                0.202    0.049    4.100    0.000    0.202    0.340
   .oh                2.376    0.241    9.863    0.000    2.376    0.647
   .es1               1.164    0.116   10.010    0.000    1.164    0.310
   .es2               0.728    0.101    7.221    0.000    0.728    0.193
   .es3               1.732    0.233    7.432    0.000    1.732    0.200
   .lot1              0.949    0.125    7.596    0.000    0.949    0.480
   .lot2              1.479    0.135   10.943    0.000    1.479    0.712
   .lot3              1.413    0.152    9.273    0.000    1.413    0.578
   .stereotype        0.017    0.005    3.511    0.000    0.517    0.517
   .health            0.198    0.049    4.013    0.000    0.506    0.506
    esteem            2.592    0.287    9.030    0.000    1.000    1.000
    optimism          1.030    0.168    6.112    0.000    1.000    1.000
```

Table 3.11b. *Local indices of the general structural model (Figure 3.7) (contd.)*

Another result that deserves our attention is that related to the R^2 of the two endogenous latent variables. Table 3.11c presents the values for this. We can see here that the portion of the variance of aging self-stereotypes that can be attributed to the set of predictive variables is 48% ($R^2 = 0.483$), while that of physical health is around 49% ($R^2 = 0.494$). The remainder can, clearly, be imputed to all the variables missing from the model.

```
R-Square:
                        Estimate
    as1                    0.191
    as2                    0.476
    as3                    0.304
    as4                    0.500
    as5                    0.208
    sh                     0.660
    oh                     0.353
    es1                    0.690
    es2                    0.807
    es3                    0.800
    lot1                   0.520
    lot2                   0.288
    lot3                   0.422
    stereotype             0.483
    health                 0.494
```

Table 3.11c. *Local indices (here R^2) of the model in Figure 3.7 (contd.).*

These results are, however, incomplete as they tell us nothing about the indirect effects of psychological resources on physical health through the self-perceptions related to aging. As we have seen, one of the shortcomings of lavaan is that it does not automatically calculate these effects. For this, it must be made clear in the specification of the model as follows.

STEP 2. Specification of the general structural model with indirect effect.

```
model.SPE <- '

# Measurement models

esteem =~ es1 + es2 + es3
optimism =~ lot1 + lot2 + lot3
stereotype =~ as1 + as2 + as3 + as4 + as5
health =~ sh + oh

# Structural model

stereotype ~ a*esteem + b*optimism
health ~ c*stereotype

# Indirect effect

indirect:= a*b*c'
```

The standardized indirect effect, displayed in Table 3.11, is 0.103, which is statistically significant (p = 0.000). This value is simply the product of three standardized effects: – 0.283* – 0.515 *0.703 (that of self-esteem on the self-stereotypes [a], that of optimism on self-stereotypes [b], and that of self-stereotypes on health [c]). Be aware that lavaan uses the Sobel test [SOB 82] to estimate the significance (*z*-value) of an indirect effect. The bootstrap procedure can also be used.

```
Defined Parameters:
                   Estimate  Std.Err  z-value  P(>|z|)   Std.lv  Std.all
    indirect          0.007    0.002    3.965    0.000    0.103    0.103
```

Table 3.11d. *Results of the indirect effects in the model in Figure 3.7 (contd.)*

We cannot ignore the fact that these results can be appreciated differently; they nevertheless confirm the theoretical representation proposed by the authors. In effect: their initial model is satisfactory not only at the statistical level, in terms of the fit with data, but also at the theoretical level, in terms of plausibility or likelihood. This model may be seen as a certain approximation of reality, even if it only partially explains the phenomenon of the role of aging self-stereotypes among the elderly. Moreover, it perhaps isn't the only model that can adjust the data. For example, an alternative model could be proposed. This is a model where health will play the mediator role while the aging self-stereotypes will play the role of the ultimate endogenous variable. We will then be in the presence of equivalent models that will be difficult to separate. The reader can refer to Bentler and Satorra [BEN 10b] for the technical aspects related to equivalent models, and can refer to Hershberger and Marcoulides [HER 13] for more on the selection criteria between these models.

3.6. Hybrid models

By hybrid model, we mean here any model that combines measured and latent variables as predictive variables, both exogenous as well as endogenous. Despite their limitations, these models translate the great flexibility of structural equation modeling. The MIMIC model (Figure 3.15) is in itself a hybrid model. The model presented in Figure 3.19 is more complex than an MIMIC model since it encloses two latent variables, including one that plays a mediator role between manifest exogenous variables (i.e. "self-esteem" and "optimism") and another ultimate endogenous latent variable (i.e. "health"). Taking the same conceptual model that was used to illustrate path models and general structural models (Figures 3.5 and 3.7) this is simply a mix of them. This is a hybrid model, described by Bentler [BEN 95] as a non-standard model.

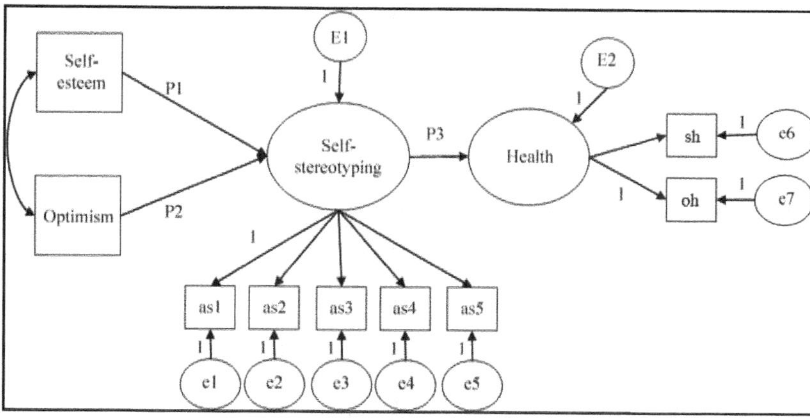

Figure 3.19. *Hybrid structural model*

3.7. Measure with a single-item indicator

Excessively minimalist version of multi-item scales, single-item scales seem to be breezing ahead. An example of a measure of perceived health with a single item whose success cannot be denied: "in general, how would you rate your health?" accompanied by a response scale with 5 (or 7) points ranging from "very bad" to "excellent".

We are not interested here in the psychometric properties of single-item measures, that is the reliability and validity of the scores obtained through single-item scales (for more on this, see, among others, [BER 07] or [PET 13]). Our interest here is more on how we can use this within a general structural model and how we can specify the particularities with lavaan.

There are two ways of integrating these measures in a general structural model. The first, as we have just seen, consists of specifying a hybrid model whose scores on single-item scales are used as measured variables, both exogenous as well as endogenous. The second consists of converting the single-item measure into a latent variable. To do this, there are two options that are necessary for the identification of such a latent variable.

Let us take the example of a construct, η (for example, perceived health) measured by a single item, X; this measurement model is translated by the following equation:

$$X = \lambda\eta + \varepsilon \qquad [3.8]$$

where:

– λ is the loading of X on the latent factor η;

– ε is the measurement error associated with the item X.

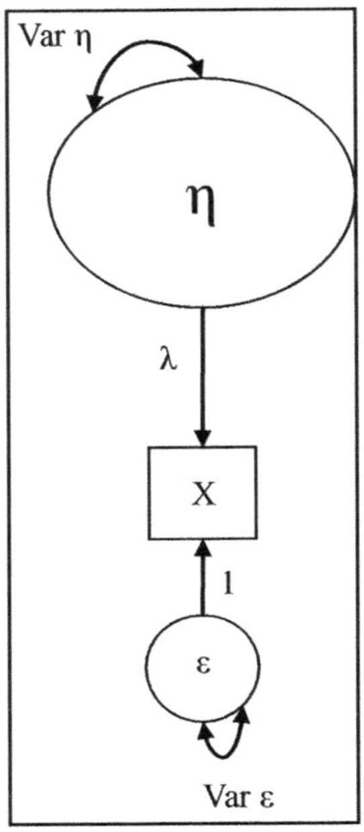

Figure 3.20. *Measurement model with a single item/indicator*

This measurement model, graphically illustrated in Figure 3.20, is not identified. It contains three parameters free to be estimated. Namely: Var η (the variance of the latent variable, which is equal to 1.00 in the case of standardization), (the factor loading of the item) and Var ε (the variance of the measurement error). And as we have only a single piece of information, namely, the variance of X (Var X), the model is then under-identified. In order to identify it (or rather, just-identify, here) and make it fit to be estimated, it must be posited that the parameter relating the unique indicator to its latent variable is equal to 1.00 (= 1.00), and its measurement error must be set to zero (Var ε = 0.00). Thus, the measured variable is considered to have no error and the latent variable (η) is considered to be a dummy variable.

Such an assumption, whose retranscription in lavaan syntax is shown below, is plausible for certain measures such as size or age (when this is duly verified). However, this is much less so when the single item is the indicator of a psychological construct such as perceived health or life satisfaction.

STEP 2. Specification of the single-item measurement model, Figure 3.20 with zero error variance.

```
model.SPE <- 'η =~ 1*X
        X ~~ 0*X' # Variance of the error associated
with X set to zero (Var ε = 0).
```

The best hypothesis in this last case: it is possible to take into account the measurement error of an item used as the unique representative of a construct by setting *a priori* its error variance, based on its assumed and proven reliability (test-retest reliability, for example) and doing this in the following manner [JÖR 89]:

$$(1 - r_{xx})*s^2 \quad [3.9]$$

where:

– r_{xx} = the recognized reliability of the single item;

– s^2 = the variance of the single item.

For example, if the variance of the single indicator X is equal to 92.78 and the test-retest reliability is estimated to be 0.70, we can set its error variance to (1 - 0.70) (92.78) = 27.83 and transcribe it in syntax lavaan as follows:

STEP 2. Specification of the single-item measure, Figure 3.20 with non-zero error variance.

```
model.SPE <- 'η =~ 1*X
              X ~~ 27,83*X'  # error variance
associated with X fixed, a priori (here Var ε = 27.83).
```

In both cases, the model is saturated, that is, just-identified (*df*=0) and can find its place in a general structural model that brings various latent variables into relation. Nonetheless, the second option seems closer to the spirit of SEM, where one of the objectives is take into account the measurement errors of the latent variables within the model that is being tested.

3.8. General structural model including single-item latent variables with a single indicator

Let us return to the hybrid model in Figure 3.19 and convert the two exogenous variables into single-item latent variables. The result is a non-hybrid structural model illustrated in Figure 3.21. Given that there is no difference at the psychometric level between an observed variable as used in the model in Figure 3.19 and a single-item latent variable whose error variance was declared to be zero, are they then interchangeable?

The response is in the affirmative, as the latent variable here is, in a way, a dummy (mute) variable. What remains to be seen is whether the results of the two models are equivalent. The response here is also in the affirmative. First, a hybrid model is no more and no less exposed than other models to identification problems. Next, a hybrid model preserves the same number of degrees of freedom as its non-hybrid counterpart. Finally, the estimation of the parameters will only be slightly affected, if at all, by the hybridization.

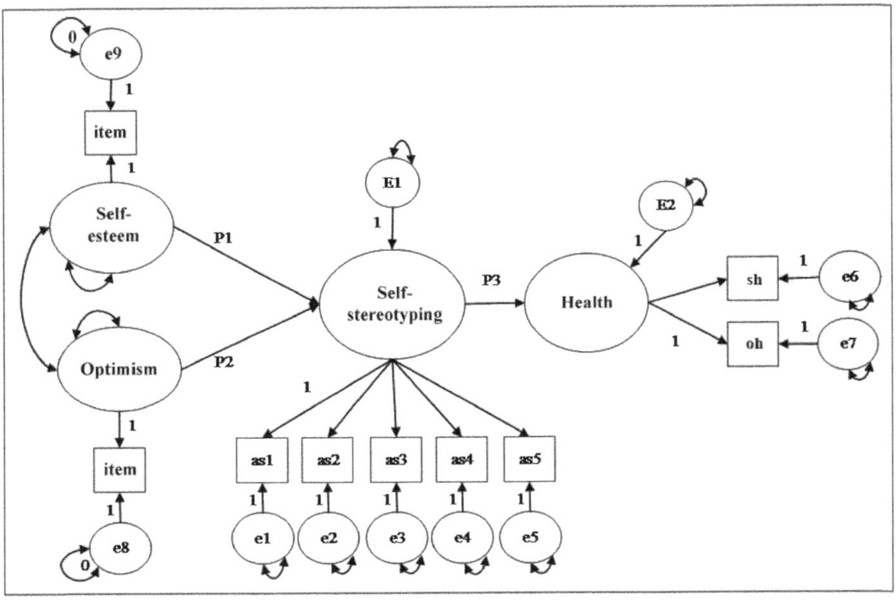

Figure 3.21. *Structural model including latent variables with a single item (the error variance of each single item is set to zero).*

3.9. Conclusion

At the end of this chapter, it seems useful to summarize the procedure leading to the validation of a model using structural equations modeling.

1- The starting point is the development of a conceptual representation that, based on one or more theories, brings into play certain relationships between a set of constructs. The diagram is useful in formalizing this conceptual representation. In effect, the confirmatory nature of the procedure requires that the theoretical specification of the model precedes the collection of data and not the other way around. This step is crucial, as it is responsible for the choice of measures of the constructs used in the specified model. Such a choice is important because the fit of the model depends on it, as well as the choice of the number of indicators for each latent variable. This is also where the rule of parsimony comes in: there is no point in complicating a model with the sole objective of achieving a statistical fit. Similarly, it is absurd to convert a recursive model into a non-recursive model to camouflage theoretical doubts concerning the predictive relationships. It is therefore preferable to propose several alternative models so that, finally, the one that best fits the data can be chosen. Furthermore, it is also recommended that models that are equivalent to the model specified by the researcher be mentioned. Models that differ in terms of the predictive links that they hypothesize substantially between the variables studied, but which generate identical estimates, that is the same model-implied variance-covariance matrix, the same number of degrees of freedom, and the same goodness-of-fit indices [HER 13] are said to be equivalent. One of the proposed solutions in order to limit the number of equivalent models is to specify, *a priori*, hypothetical models that include variables (for example, instrumental (or control)) that are not related to all the variables studied. This is to avoid specifying hypothetical models where everything would be related to everything else. A model is chosen, first of all, based on theoretical and conceptual considerations. Certain statistical indices, which are difficult to implement, have been suggested in order to distinguish between equivalent models [RAY 01]. Others recommend comparing the R^2 of the endogenous variables in order to distinguish between the equivalent models (for more on this subject, see [WIL 12]).

2- Going on to estimation in two steps: first the measurement model and then the structural model.

3- Evaluating the measurement model based on the Overall goodness-of-fit indices and local fit indices: if the results are satisfactory, we can then move on to step 5.

4- Improving the measurement model by taking into account the modification indices and theoretical considerations: after each modification, we must respecify the model and go back to step 3. There is no point in statistical obstinacy, to use any means to get the measurement model to fit the data.

5- Evaluation the structural model by examining the overall goodness-of-fit indices as well as the local fit indices : if the solution is proper and satisfactory, we can then move on to step 7.

6- Improving the structural model, if necessary, by using modification indices. After each respecification, which must be theoretically justified, it is imperative to return to step 5.

7- Proceeding to the interpretation of the results: however, when *post-hoc modifications* have been made to achieve a satisfactory fit, it is recommended to consider a cross-validation of the model.

It is necessary to be aware that even a perfect fit does not necessarily prove the veracity of a model. As long as the unique character of a model (i.e. the absence of any equivalent model) is not established, its fit with the data can only be considered a good approximation of reality.

We conclude this chapter by going back to Figure 3.1, which served as an introduction to this chapter. We complement it (Figure 3.22) by highlighting the "Model estimation" – "Evaluation of the solution or model evaluation" – "Model modification" loop in order to give a more detailed synthetic overview of the steps in structural equation modeling (see [GRA 05]).

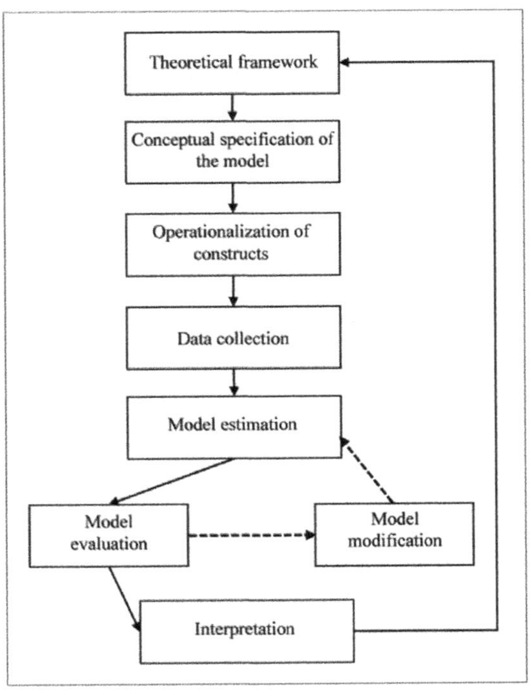

Figure 3.22. *The main steps in application of SEM*

3.10. Further reading

For readers who may wish to study the concepts discussed in this chapter in greater detail, we refer the following books:

BOLLEN K.A., *Structural Equations with Latent Variables*, Wiley, New York, 1989.

BOLLEN K.A., LONG J.S., *Testing Structural Equation Models*, Sage Publications, New York, 1993.

BROWN T.A., *Confirmatory Factor Analysis for Applied Research*, 2nd edition, Guilford Press, New York, 2015.

KLINE R.B., *Principles and Practice of Structural Equation Modeling*, 4th edition, Guilford Press, New York, 2016.

LOEHLIN J.C., *Latent Variable Models: An Introduction to Factor, Path, and Structural Equation Analysis*, 4th edition, Lawrence Erlbaum Associates Publishers, Mahwah, 2004.

MARUYAMA M.G., *Basics of Structural Equation Modeling*, Sage Publications, New York, 1998.

4

Advanced Topics: Principles and Applications

The extensions of structural equation modeling are multiple, and its applications increasingly numerous and varied. We should immediately recognize that it is impossible to draw up an inventory. The growing interest that has not ceased to be shown in these methods has led to a considerable broadening of their application in various scientific domains and disciplines. In addition to complex longitudinal hierarchical data (multilevel; [MUT 02a]), they are also applied now to experimental data [BLA 09, PLO 04], or indeed to both simultaneously [MCA 07], to neuroimaging data [MCI 12] and clinical medical data [ZHA 17]. This is not at all trivial, above all when it is known that these methods were destined instead for domains where the possibilities for direct experimental verifications could not, for material and/or deontological reasons, be envisaged, and in which it is necessary to proceed otherwise to fully consider relationships between the different variables studied. Structural equation modeling was, undoubtedly, the right tool.

Here, we will merely draw out the simplest extensions; so simple that they sometimes cease to be considered as extensions. And their simplicity is only as great as their immense flexibility. Judge for yourself, using the different examples throughout this chapter. Without forgetting that today's extensions will be quickly and fortunately obsolete tomorrow.

4.1. Multigroup analysis

Comparing several groups from different aspects is a common procedure in scientific research (for example analysis of variance [ANOVA], multivariate analysis of variance [MANOVA], multivariate analysis of covariance [MANCOVA], repeated measures analysis of variance [RANOVA]). Comparing them through the lens of one or more theoretical models is much less so, since, to our knowledge, only structural equation modeling can make this possible. The primary aim for such a comparison is to estimate a model's invariance across populations, thus making it possible to establish the validity of the theoretical construction, which the model is the representation of. The question is whether estimation of the parameters of the posited model would vary across the groups involved. Let us put this differently: multigroup analysis, also referred to as "multisample analysis", makes it possible to test a model's plausibility in different groups and is thus a fairly economical way for detecting differences between them. The fact that a model proves applicable or not to other populations makes it possible, in principle, to judge the generalizeability, or otherwise, of the hypothetical assertions it conveys. It is clear that a theoretical representation is even more worthwhile when/if it is suitable for several samples, as the worst-case scenario would be to finish with as many models as samples.

Thus formulated, this analysis would instead resemble a cross-validation procedure: a model applied to a sample will be valid for another, thus making it possible to establish an already proven model's stability and invariance. But, although this end is evidently not excluded, multigroup analysis is distinguished by the fact that estimation of the model is not obtained for each group separately and successively, but rather for all the groups analyzed simultaneously. Of course, we could be content with the first option and examine the results of each group separately without being able to state their equivalence or dissimilarity statistically. Since it is clear, for example, that the absence of differences between the individual parameters from one group to another absolutely does not mean that these parameters are equivalent. Indeed, the move from separate estimations to simultaneous consideration of all the groups involved is such a good opportunity and such flexibility is offered by structural equation modeling that it seems inconceivable to us to dispense with it/ignore it, especially as it allows us to test parameter equality across groups, and above all it provides a statistical test making it possible to decide on the comparability of these groups.

There is scarcely any need to recall that what is being compared is really a network of relationships between variables specified by the researcher. This remark calls for another, just as fundamental here: multigroup analysis necessarily requires use of the variance-covariance matrix rather than the correlation matrix. The reason for this is very simple: given that, in a correlation matrix, all the variances were fixed at 100, they bring nothing to the comparison. Although they are neither the only information available nor the only information needed for the analysis (the measured variable means are also necessary), the variances of the variables in each of the groups studied will be, for comparison, opposed. We will admit that they are subject to differences, and starting from there, that they are useful and interpretable sources of information.

Similarly, we agree on recognizing that use of a correlation matrix leads to incorrect solutions when parameter equality constraints are imposed, which is as it happens is the case in multigroup analyses, and which must be discussed later. It is preferable, here too, to analyze the variance-covariance matrix and then standardize the solution.

Multigroup analysis applies to/covers all models that fall within the scope of structural equation modeling. It applies to path models, general structural models, measurement models and latent growth models, among others. It is the measurement models that we select for this chapter as they take a considerable place in applied psychometrics. Isn't a measure's factorial structure at the heart of the definitional theory that underlines and founds this measure? We have chosen to show how multigroup analysis services the study and examination of this structure. This is called a multigroup confirmatory factor analysis (MG-CFA). It is an extension of a simple CFA, to the extent that it becomes an analysis of mean structures involving estimation of item intercepts and latent variable means, and their intergroup comparison. An indicator intercept (for example an item), as has already been seen, refers to its estimated mean when the factor on which it depends is equal to zero (see [KLI 16] for mathematical details relating to indicator intercepts).

Recall that a construct's structure refers to different sources of variation in the responses to the items that define and represent it. This structure is one-dimensional where there is only one source of variation identified; it is, however, multidimensional when there are several. Each source of variation is a dimension (factor). Correlated between one another, these dimensions give body to a multidimensional oblique solution.

Testing measurement and structural invariance participates in the intra-construct validation process which seeks to show that a construct's structure and metrics are independent of the population to which it applies. This invariance is posed even more sharply as the comparison concerns culturally different groups. Do items on a questionnaire measure the same construct with the different groups and sub-groups to which it is administered?

Measurement invariance concerns indicator characteristics (for example items), that is their factor loadings, their intercepts and their measurement error variances, whereas the structural invariance, which is not necessary for the quality of a measure, deals with the characteristics of the model's latent variables, that is their variances, covariances and means. But there are several degrees of invariance in a measure [REN 98], for which the terminology varies from one author to another:

– a configural invariance which refers to intergroup equivalence of the factor structure (i.e. equal form). Testing such an invariance means trying to answer the question of how we can know if the groups at hand share the same general factorial pattern of the measure (i.e. the same number of factors and the same factorial pattern). This means the least restrictive model, in which no equality constraint has been imposed on the groups. We only formulate the hypothesis of the existence of the same numbers of factors and the same loading patterns in the groups compared. This model is doubly important: it is the baseline model on which the remainder of the tests depend, as if this model, the least restrictive one, fails to adjust the data there is scarcely any chance for less restrictive models to reach it;

– a weak invariance, also referred to as metric invariance, which assumes intergroup equality of factor loadings. Testing this invariance means trying to answer the question of whether the groups at hand display the same factor loadings. Such an invariance, which is expressed by intergroup equivalence of relationships between the items and the latent variables on which they depend, indicates that the different groups share the same significance for the items composing a measure;

– a strong invariance, also known as scalar invariance, which assumes intergroup equivalence as much for factor loadings as for those of items' intercepts. This condition will integrate the means of the items measured in the analysis (hence the need to introduce them at the same time as the variance-covariance matrix in the absence of raw data). Such a condition, cumulating in the configural invariance, that of the factor loadings, but also those of the intercepts, indicates that, for a given value of the latent variable, the scores for items are assumed to be identical in the different groups at hand. In other words, a unit variation at the level of the latent variable is linked to an identical change at the level of the scores for items in the groups at hand. Thus, comparison of scores to these groups' construct takes its full meaning here, and if there is a difference, it means a real difference at the level of the construct and not a difference that can be imputed to the way it is measured;

– a strict invariance that assumes intergroup equality at once in items' factor loadings, in their intercept and their measurement error variance. This fairly demanding condition is rarely reached, as it assumes that the measurement reliability is perfect and identical among the groups at hand and that the measurement is not affected by any bias. Each item's variance is thus explained identically in each group at hand, and the construct is measured identically in each group. The scores' intergroup differences in measure convey the real differences relating to the construct measured.

An MG-CFA obeys a sequence wherein passing from one stage to another requires successfully passing through the previous step: (1) applying a simple CFA to each group at hand separately (i.e. there will be as many CFAs as groups), as it requires that the posited model (i.e. whose invariance we wish to test) should be tolerated by each group at hand; (2) testing the configural invariance which consists of evaluating simultaneously and without any equality constraint the model that has just been estimated separately; (3) testing the weak invariance which consists of constraining all the factor loadings in the groups to equality (or in some cases only some factor loadings, and we speak here of "partial invariance". We will return to this later); (4) testing the strong invariance which consists of raising the previous level of equivalence by adding to it an intergroup equality constraint involving items' intercepts; (5) testing, if necessary, (and desirable) the strict invariance which consists of adding to all the previous constraints on intergroup equality of measurement error variances; (6) testing the equivalence across groups of latent variable means; (7) testing the equivalence across groups of latent variable variances, and (8) the intergroup invariance of covariances between latent factors (where there is a multidimensional model).

The first five steps corresponding to evaluation of the measurement invariance are most commonly used in the context of MG-CFA analyses. In addition, satisfying the first four stages is more than enough to admit a measurement invariance. Stages 6, 7 and 8 correspond to tests of the hypothesis of structural invariance.

The shift from one step to another is subject to a test, which involves examining the statistical significance of the difference in the values of χ^2 for two nested models ($\Delta\chi^2$), depending on the difference in the degrees of freedom (Δdf). We compare the χ^2 of the weak invariance to that of the configural invariance. We then compare the χ^2 of the strong invariance to that of the weak invariance. Finally, we compare χ^2 of the strict invariance to that of the strong invariance. When the delta-χ^2 ($\Delta\chi^2$) is statistically significant (to at least $p < 0.05$), it is legitimate to conclude that the additional constraints imposed on the model (for example, intergroup equality of the items' intercepts) have deteriorated its overall fit so that there is no intergroup invariance (configural, scalar or strict).

Although $\Delta\chi^2$ remains the most commonly used test for judging the probability of the invariance, differences relating to some fit indices such as CFI (ΔCFI = delta-CFI) and RMSEA (ΔRMSEA = delta-RMSEA) have also been suggested.

A ΔCFI ≤ 0.01 indicates that the null hypothesis of invariance cannot be rejected. A differential higher than 0.01 is indicative of a deterioration in the model that has been subject to the new equality constraints, thus indicating the improbability of the invariance. In the same vein, a minimal ΔRMSEA is indicative of the invariance's plausibility. Other tests look at the RMSEAs' confidence interval. If the values of the two models in competition fall in the same interval, it is assumed that there has not been any deterioration in fit of the model subject to additional equality constraints.

As an illustration, we take here the dispositional hope scale whose factorial structure was studied in Chapter 3 of this book (Figure 3.11, Chapter 3). We subject this structure to a MG-CFA to test its measurement invariance across gender.

Once the hypothetical factorial structure is specified, here as it happens it is two-factor (Figure 3.11), and before proceeding to the hypothesis tests on the equivalence across groups, it is necessary to verify separately the model fit for each of the two sexes. We agree that it is useless to compare groups from the angle of a model when this is not acceptable in both groups. It is therefore vital, first to test the baseline model, which will then be subject to equivalence tests across groups. These tests are carried out using simultaneous analyses of the model, applied to several groups. It is however usual to test, successively a series of equivalence hypotheses, the logic for which we have just presented: to go from less equivalence to more equivalence across groups, on the parameters forming a measurement model. We will now reveal these stages, step by step.

4.1.1. *The steps of MG-CFA*

4.1.1.1. *MG-CFA stage one: testing the CFA model for each group separately*

We will first specify the model's parameters in lavaan syntax. The box below provides the details of this specification. We recall here, and this will be the case throughout this section, that the "Steps" refer to the main stages of using lavaan detailed in the first part of this book. We also recall that the comments that follow the hashtag (#) are not part of the commands, but explain them.

STEP 1. Importing data and creating subsamples.

#1. Create a subsample with data on women.

```
women <- subset (BASE, sex == "1")
```

#2. Create a subsample with data on men.

```
men <- subset (BASE, sex == "2")
```

STEP 2. Specifying the two-factor model of the hope scale (Figure 3.11).

```
model.SPE <- 'pathways =~ hop.p1 +  hop.p3 + hop.p4
+ hop.p5
agency =~ hop.a2 + hop.a6 + hop.a7 + hop.a8'
```

STEP 3. Model estimation in women subsample.

```
model.EST <- cfa (model.SPE, data = women, estimator
= "MLR")
```

STEP 3 REPEAT. Model estimation in men subsample.

```
model.EST <- cfa (model.SPE, data = men, estimator
= "MLR")
```

The stage of testing the CFA model separately in each group requires either the use of a dataset specific to each group, or a partitioning of the dataset containing the shared data to create two sub-samples of it, one regrouping data for women and the other data for men. If we opt for the second solution, sections #1 and #2 show the way we must proceed to achieve it. We detail the elements in section #1: the notification "women" is the name chosen for the female sub-sample initially coded "1" in our group variable "sex"; "subset" is a function of R making it possible to create a sub-sample from our "BASE" dataset imported in step 1. "sex == "1"" makes it possible to create the sub-sample regrouping all the individuals coded "1" with the variable "sex". It is the same for men, for whom the sub-sample regroups all the individuals coded "2" with the variable "sex". Each of these sub-samples will be used in the place of "BASE" to estimate separately the model for each group, as the commands figuring on steps 3 and 3bis in the box below show. The estimator MLR has been retained, as this was the case for the two-factor model estimated previously (see Figure 3.11).

We will spare the reader the details of the evaluation of the two solutions. It will only be said that as the model's plausibility for each group is confirmed, and the modifications consequently useless, we can therefore move to the following step. It will be recalled above all that the number of degrees of freedom for each group was 19 (see the results of the model represented by Figure 3.11, Chapter 3).

4.1.1.2. *MG-CFA stage 2: configural invariance test*

The box below synthesizes the commands for each stage. It will be noted that step 2 of the two-factor models' specification remains unchanged. The essential part is found in step 3, the stage specifying the model's estimation.

We see first that the estimation covers all the data (data = BASE). We then found, and this is absolutely not a mere detail, a fundamental indication: "group = "sex"" without which MG-CFA will not take place, meaning here that it is a multigroup analysis, and that it is the variable "sex" in our dataset (BASE) that will differentiate the groups in competition.

STEP 2. Specifying the two-factor model of the hope scale (Figure 3.11, Chapter 3).

```
model.SPE <- 'pathways =~ hop.p1 + hop.p3 + hop.p4
+ hop.p5
agency =~ hop.a2 + hop.a6 + hop.a7 + hop.a8'
```

STEP 3. Estimating the measurement configural invariance.

```
configural <- cfa (model.SPE, data = BASE, group
= "sex", estimator = "MLR")
```

STEP 4. Retrieving the results including the modification indices.

```
summary (configural, fit.meaures = T)
```

This step is essential, if it passes the test for moving stages, in other words if the configural invariance is seen to be tolerated by the data, we can then move to the following stage and thus continue the MG-CFA. As it happens, this was the case here, the goodness-of fit indices show: robust-CFI = 0.972, robust-TLI = 0.959, robust-RMSEA = 0.043 [90% CI = 0.000 – 0.073]. Also, reported will be $\chi_{ML}(38) = 53.77$, $p = 0.046$ and $\chi_{robust}(38) = 48.71$, $p = 0.114$. The reader may perhaps be surprised to see the number of *df* equal to 38. We will return to this, commenting further in Table 4.3.

4.1.1.3. *MG-CFA stage three: weak invariance test hypothesis*

The weak invariance test, which is the hypothesis of equality across groups (sexes here) of factor loadings, requires the use of a new command at the estimation step: “group.equal = "loadings"”; "group.equal" means “equality across groups”, and "loadings" means that this equality constraint concerns the factor loadings. It will be noted that the two-factor model specified remains unchanged.

STEP 2. Specifying the two-factor model of the hope scale (Figure 3.11).

```
model.SPE <- 'pathways =~ hop.p1 + hop.p3 + hop.p4 +
hop.p5
agency =~ hop.a2 + hop.a6 + hop.a7 + hop.a8'
```

STEP 3. Estimating the weak measurement invariance.

```
weak <- cfa (model.SPE, data = BASE, group = "sex",
group.equal = "loadings", estimator = "MLR")
```

If, compared to the previous model (i.e. $\Delta\chi^2$, Δdf), this stage does not cause deterioration in model fit, it is therefore permissible to move to the following stage of invariance tests. We will see later how to proceed to compare the models' goodness-of-fit.

4.1.1.4. *MG-CFA stage 4: strong invariance test hypothesis*

The strong invariance test which makes the hypothesis of equality across groups both of factor loadings and indicator intercepts (estimated means) requires the command from the previous stage to be completed: group.equal = c ("loadings", "intercepts"). Thus, group equality across ("group.equal") concerns/involves the factor loadings ("loadings") as well as the indicator intercepts. The box below shows the details.

STEP 2. Specifying the two-factor model of the hope scale (Figure 3.11).

```
model.SPE <- 'pathways =~ hop.p1 + hop.p3 + hop.p4 +
hop.p5
agency =~ hop.a2 + hop.a6 + hop.a7 + hop.a8'
```

STEP 3. Estimating the strong measurement invariance.

```
strong <- cfa (model.SPE, data = BASE, group = "sex",
group.equal = c ("loadings","intercepts"), estimator =
"MLR")
```

If, compared to the previous model (i.e. $\Delta\chi^2$, Δdf), this stage will not degrade the model's fit, we are then authorized to move to the following stage of invariance tests. Comparison of the models' fit will be addressed further.

4.1.1.5. *MG-CFA stage 5: strict invariance test hypothesis*

An additional equality constraint, that relating to the indicators' measurement error variances "residuals", is added to the previous ones, thus resulting in reflecting the strict invariance hypothesis: group.equal = c ("loadings", "intercepts", "residuals").

STEP 2. Specifying the two-factor model of the hope scale (Figure 3.11).

```
model.SPE <- 'pathways =~ hop.p1 + hop.p3 + hop.p4 +
hop.p5
agency =~ hop.a2 + hop.a6 + hop.a7 + hop.a8'
```

STEP 3. Estimating the strict measurement invariance.

```
strict <- cfa (model.SPE, data = BASE, group = "sex",
group.equal = c ( "loadings", "intercepts",
"residual"), estimator = "MLR")
```

4.1.1.6. *MG-CFA stages 6–7–8: tests for structural invariance hypotheses*

These stages form part of the tests for the hypothesis of structural invariance whose requirements have just been added to those required for the measurement invariance. Successively, we move from one stage to another. Let's look at them one by one:

– at stage 6, the equality of latent variable means is added to the previous constraints: group.equal = c ("loadings", "intercepts", "residuals", "means");

– at stage 7, the equality of latent variable variances is added: group.equal = c ("loadings", "intercepts", "residuals", "means", "lv.variances");

– at stage 8, the equality of the covariances between the latent variables is added: group.equal = c ("loadings", "intercepts", "residuals", "means", "lv.variances", "lv.covariances").

It will be specified that "means" refers to the latent variable means, "lv.variances" indicates latent variable variances the latent variable variances (lv = latent variable) and "lv.covariances" designates latent variable correlations/covariances.

It will also be remembered that we move from one stage to another when the added equality constraints do not cause the model to deteriorate.

STEP 2. Specifying the two-factor model of the hope scale (Figure 3.11).

```
model.SPE <- 'pathways =~ hop.p1 + hop.p3 + hop.p4 +
hop.p5
agency =~ hop.a2 + hop.a6 + hop.a7 + hop.a8'
```

STEP 3. Estimating the measurement and structural invariance.

```
structural <- cfa (model.SPE, data = BASE, group =
"sex", group.equal = c ("loadings", "intercepts",
"residuals", "means", "lv.variances",
"lv.covariances"), estimator = "MLR")
```

4.1.2. *Model solutions and model comparison tests*

Here, we will merely examine the measurement invariance (configural, weak, strong and strict). Two options are offered to the user to obtain solutions for different estimated invariance models and to perform model fit comparison.

The first involves turning to the usual "summary ( )" function to obtain results for each model estimated separately (see sections #1 to #4 in the box below) and to then use the "anova" function (section #5) for the for model comparison tests; a glimpse of its results is shown in Table 4.1.

#1. Estimation and solution of the configural invariance model.

```
> configural <- cfa (model.SPE, data = BASE, group =
"sex", estimator = "MLR")
> summary (configural, fit.measures = TRUE)
```

#2. Estimation and solution of the weak invariance.

```
> weak <- cfa (model.SPE, data = BASE, group = "sex",
group.equal = "loadings", estimator = "MLR")
> summary (weak, fit.measures = TRUE)
```

#3. Estimation and solution of the strong invariance.

```
> strong <- cfa (model.SPE, data = BASE, group = "sex",
group.equal = c ("loadings", "intercepts"), estimator =
"MLR")
> summary (strong, fit.measures = TRUE)
```

#4. Estimation and solution of the strict invariance.

```
> strict <- cfa (model.SPE, data = BASE, group = "sex",
group.equal = c ("loadings", "intercepts",
"residuals"), estimator = "MLR")> summary (strict,
fit.measures = TRUE)
```

#5. Comparison of models (the models to be compared, at least two, are in parentheses).

```
> anova (configural, weak, strong, strict)
```

Scaled Chi Square Difference Test (method = "satorra.bentler.2001")

	Df	AIC	BIC	Chisq	Chisq diff	Df diff	Pr(>Chisq)
configural	38	8027.1	8213.9	53.777			
weak	44	8017.1	8181.5	55.816	1.7288	6	0.9429
strong	50	8014.5	8156.4	65.139	9.0995	6	0.1681
strict	58	8009.6	8121.7	76.316	6.5511	8	0.5857

Table 4.1. *Results of the "anova ( )" function*

The first column in Table 4.1 lists the models subject to the comparison test. Column "Df" displays the number of their respective degrees of freedom; those of AIC and BIC offer the values of two goodness-of-fit indices used to decide between the models in competition (the better model fit, the weaker the values of these indices); column "Chisq" displays the values of χ^2; column "Chisqdiff" shows $\Delta\chi^2$ (χ^2 of the model – χ^2 of the previous model) along with Δdf given to the column "Dfdiff"; the last column, which is far from being the most interesting, shows the p-value of $\Delta\chi^2$ so as to be able to judge the statistical significance. And it will be noted that despite the use of the MLR estimator, the χ^2 difference test concerns the values of χ^2_{ML}. These differences are corrected following the Satorra-Bentler formula [SAT 01]. Indeed, the value 1.72, which quantifies the difference between the χ^2 of the "weak" model and that of the "configural" model does not correspond exactly to 55.81 – 53.77 (= 2.04). This difference of 2.04 has been scaled to obtain 1.72. Finally, it will be noted that lavaan does not offer ΔCFI and ΔRMSEA.

From the results, it can be seen that the equality constraints imposed to test a weak invariance have not affected the model fit relative to the configural solution. Indeed, the differential of their respective χ^2 ($\Delta\chi^2$ = 1.72), compared to that corresponding to their respective degrees of freedom (Δdf = 6) has proven not to be significant (p = 0.942). This suggests that the hypothesis of factor loading equality in the two groups is plausible. Such a result means we can test the hypothesis of a much stronger invariance and ask whether or not this will deteriorate the model's fit. To know this, it is enough to compare the strong invariance model to the weak invariance model. The differential of their respective χ^2 ($\Delta\chi^2$ = 9.09), compared to that corresponding to their respective degrees of freedom (Δdf = 6), has revealed itself not to be significant (p = 0.168). Thus, the hypothesis of a strong invariance reflecting factor loading aquality as well as equality of indicators' intercepts, is plaisible. If we now compare the overall fit of the strict invariance model with that of the strong invariance model (χ^2 (58) = 76.31) with that of strong invariance (χ^2(50) = 65.13), it will be noted that the additional equality constraints that affect the first relative to the second have not affected its overall fit, since $\Delta\chi^2$ scaled = 6.55, with a Δdf = 8, is not statistically significant (p = 0.585).

The second option that the user of lavaan and R has to obtain model comparison tests is to use the "measurementInvariance ( )", function which has unfortunately ceased to be an integrated lavaan function to become a "semTools" function. This function offers a shortcut making it possible to estimate automatically the different invariance models (configural, weak, strong, LV means) and to compare their goodness-of-fit. It will be noted, and this is very odd, that the strict invariance is not considered automatically, and to perform goodness-of-fit comparison. The box below provides the necessary details.

STEP 2. Specifying the two-factor model of the hope scale (Figure 3.11).

```
> model.SPE <- 'pathways =~ hop.p1 + hop.p3 + hop.p4 +
hop.p5
agency =~ hop.a2 + hop.a6 + hop.a7 + hop.a8'
```

#1. Open the semTools package.

```
> library (semTools)
```

STEP 3. Comparison of invariance models using semTools' "measurement Invariance".

```
> models <- measurementInvariance (model.SPE, data =
BASE, group = "sex", strict = TRUE, estimator = "MLR")
```

If necessary, obtain the χ^2 of each invariance model separately.

```
> models [[1]]
> models [[2]]
> models [[3]]
> models [[4]]
> models [[5]]
```

Let us stop for a moment at STEP 3, as the specified two-factor model remains unchanged. The object is arbitrarily called "models"; "measurementInvariance" represents the semTools function that carries out the invariance tests; between parentheses, there are the options: "model.SPE" is the name of the model specified to be tested for invariance, "BASE" is the name of the dataset needed to estimate the model, "group = "sex"" specifies the group variable and finally, "strict = TRUE" is a request to test the strict invariance .

The "measurementInvariance" function also displays the advantage of offering ΔCFIs and ΔRMSEAs. Table 4.2 shows a glimpse. At the top of this table, the five invariance models are recalled: (1) "configural", (2) "loadings" = weak invariance, (3) "intercepts" = strong invariance, (4) "residuals" = strict invariance, and (5) "means" = structural invariance of the means of the latent variables. In the middle of the table, we find the χ^2 difference test, and at the bottom of the table, the tests for the differences in CFI and RMSEAs.

There, you can read that the only noticeable deterioration in the fit of the models in competition is caused by the equality constraint of latent means carried by model 5 (fit.means). Indeed, the $\Delta\chi^2$ of models 4 and 5, which is equal to 17.1946 with a Δdf equal to 2, is revealed to be statistically significant ($p = 0.0001846$), indicating that the equality across groups of latent variable means "pathways" and "agency" is unlikely. The ΔCFIs and ΔRMSEAs corroborate this result. It is clear that CFI of model 5 (0.949) shows substantial drop compared to the previous model ("fit.residuals"), which will serve it as a comparison model (0.984). It is the same for the RMSEA, which has seen its value increase, moving from 0.027 to 0.046 due to the intergroup equality constraint of the latent variable means.

The "models [[]]" inventory makes it possible to obtained separately the χ^2 of each invariance model, of which Table 4.3 offers only those, as an illustration, of the configural model[[1]].

```
Measurement invariance models:

Model 1 : fit.configural
Model 2 : fit.loadings
Model 3 : fit.intercepts
Model 4 : fit.residuals
Model 5 : fit.means

Scaled Chi Square Difference Test (method = "satorra.bentler.2001")

               Df    AIC    BIC  Chisq Chisq diff Df diff Pr(>Chisq)
fit.configural 38 8027.1 8213.9 53.777
fit.loadings   44 8017.1 8181.5 55.816     1.7288       6  0.9428767
fit.intercepts 50 8014.5 8156.4 65.139     9.0995       6  0.1680597
fit.residuals  58 8009.6 8121.7 76.316     6.5511       8  0.5857463
fit.means      60 8023.8 8128.4 94.492    17.1946       2  0.0001846 ***
---
Signif. codes:  0 '***' 0.001 '**' 0.01 '*' 0.05 '.' 0.1 ' ' 1


Fit measures:

               cfi.scaled rmsea.scaled cfi.scaled.delta rmsea.scaled.delta
fit.configural      0.972        0.043               NA                 NA
fit.loadings        0.984        0.030            0.012              0.013
fit.intercepts      0.977        0.034            0.008              0.004
fit.residuals       0.984        0.027            0.007              0.008
fit.means           0.949        0.046            0.035              0.020
```

Table 4.2. *Results obtained with the "measurementInvariance" function*

```
> models[[1]]
lavaan (0.5-23.1097) converged normally after  52 iterations

  Number of observations per group
  1                                              162
  2                                              148

  Estimator                                       ML       Robust
  Minimum Function Test Statistic             53.777       48.718
  Degrees of freedom                              38           38
  P-value (Chi-square)                         0.046        0.114
  Scaling correction factor                                 1.104
    for the Yuan-Bentler correction

Chi-square for each group:

  1                                           24.242       21.961
  2                                           29.535       26.757
```

Table 4.3. *Example of the results of χ² for each group for each invariance model (Model [[1]] = configural model; 1 = women group ; 2 = men group)*

Table 4.3 shows the global χ^2 as well as each group's χ^2 (1 = women, 2 = men) for the model [[1]], in this case the configural model. It will be specified that the value of χ^2 is equal to the sum of χ^2 of the groups at hand ($\chi^2_{ML} = 53.77 = 24.24 + 29.53$). It is the same for the number of degrees of freedom ($df = 19 + 19 = 38$).

These results, leading to a strict invariance in the measure of dispositional hope scale between men and women, enable us to conclude that there is no gender bias affecting answers to this scale. This scale therefore evaluates the same construct, in the same way in both the groups. This absolutely does not mean that there are no differences between men and women relating to their dispositional hope. This means that if there are differences, they can be imputed to real differences at the level of the construct itself and not to gender biases affecting the measure. This scale can therefore be called gender-free, that is "gender-neutral scale".

4.1.3. *Total invariance versus partial invariance*

A two-fold question is permissible here. If weak invariance, considered as the minimal condition of measurement invariance, is not confirmed, must we then abandon the entire procedure underway, or indeed can the non-invariant items be detected, and so settle with a partial invariance, tolerating the fact that some factor loadings can differ from one group to another?

Researchers like Byrne *et al.* [BYR 89] recommend such a relaxation, as it seems difficult to reach a total factorial invariance, of course on the condition that the number of non-invariant items is reasonable relative to the number of invariant items. We can in fact, by recourse to modification indices, ensure that each equality constraint imposed in the model is acceptable. These indices make it possible to detect equality constraints that seem implausible and which we can shed to improve the model's overall fit.

The usefulness of these indices, which are obtained using the option "modindices = TRUE" in "summary", is far from being negligible, above all when we wish to screen the problematic items to diagnose their effects and eventually provide solutions to them, especially in the context of cross-cultural validations of scales. However, the the goodness-of-fit indices should not exempt us from underpinning our decision to accept partial invariance conceptually.

4.1.4. *Specification of a partial invariance in lavaan syntax*

As an illustration, let us assume that we can justify (conceptually and empirically from the modification indices) removing the equality constraint imposed on an item, which we let us arbitrarily take item hop.p3.

```
> weak <- cfa (model.SPE, data = BASE, group = "sex",
group.equal = "loadings", group.partial = "pathways =~
hop.p3", estimator = "MLR")
```

In the box below, a new indication has appeared: "group.partial", indicating the presence of a partial invariance in the model subject to estimation. This partial invariance concerns item hop.p3 (group.partial= "pathways =~ hop.p3"), whose loading on the "pathways" factor has dispensed with any equality constraint, that is it is permitted to differ from one group to another.

```
> strong <- cfa (model.SPE, data = BASE, group = "sex",
group.equal = c ("loadings","intercepts"),
group.partial = c ("pathways =~ hop.p3", "hop.a6 ~1"),
estimator = "MLR")
```

In the specification above, all equality constraints across groups have been removed, to the loading of item hop.p.3 on the factor "pathways" ("pathways = ~ hop.p3") as well as to the intercept of the item hop.a6 ("hop.a6..)

4.2. Latent trait-state models

Latent trait-state models are the name for a type of model designed with the intention of examining stable and non-stable aspects of a construct over time. Stable aspects refer to trait variance while non-stable aspects refer to state variance as well as error variance [ALL 36, CAT 47, EYS 83]. Although this debate has always been lively [STE 15], there is however a consensus [FRI 86] that psychological traits are considered as relatively stable personal attributes, arising from an individual's tendencies, propensities, or even styles and ways of being, feeling and thinking in a particular situation.

Hertzog and Nesselroade [HER 87], not only highlight the fact that most psychological attributes are neither strictly traits nor strictly states, but they also draw attention to two widely shared false ideas on psychological traits and states. First, according to them, traits should not be used to mean unchangeable and genetically determined psychological attributes. Second, states are wrongly considered ephemeral and unpredictable and are unreliably evaluated (because of a confusion between stability and reliability). Kenny and Zautra [KEN 01] highlight the idea that all psychological constructs vary along a stability continuum called traitness. It remains to be seen what is part of the trait and what is part of the state when measuring a psychological construct. In other words, what does the score obtained from a measure capture: a psychological trait or rather a psychological state? We will pose this question again in another way: what part of the variance is imputable to the trait and what part is imputable to the state in a score obtained at any assessment occasion? It is this question that latent trait-state models aim to answer.

The reader will no doubt have understood that when we speak of stability, were are dealing with logitudinal data analyses. We must specify as much to begin with: these models require at least four repeat assessments of the same construct spread over time with the same participants (at least 3–4 waves of assessment). Here, we will not address attrition, which is an inherent limitation in longitudinal studies, for which recent statistical advances are making it possible to reduce biases a little in estimating the models' parameters. In Graham [GRA 09] the reader will find a fairly accessible explanation of modern statistical models for handling missing data.

This chapter suggests providing a general overview introducing two variants of latent trait-state models: (1) Kenny and Zautra's model [KEN 95, KEN 01] first called "trait-state-error model", then renamed Stable-Trait Autoregressive-Trait and State Model or STARTS; (2) Cole, Martin and Steiger's model [COL 05], known by the name of Trait-State-Occasion Model or TSO.

We will explain how they are specified and identified using lavaan, and we will comment on their results using illustrations.

4.2.1. *The STARTS model*

This model can be divided into two versions: a univariate STARTS and a multivariate STARTS, which we will present in turn.

4.2.1.1. *Univariate STARTS*

This model, of which Figure 4.1 provides a visualization in diagram form, is applied to a single observed measure of the construct (for example a scale's total score) at least four times running with an interval (time lags) between assessment points, fixed by the researcher.

This model makes the hypothesis that each time a assessment is made, the variable measured (y_1, y_2, y_3, y_j) can be a function of three independent latent variables: a time-invariant latent factor, reflecting a stable trait (ST), a time-varying factor, reflecting an autoregressive trait (ART), and finally a state factor (S), reflecting the specific measurement occasion (i.e., time-specific effects) as well as measurement error:

$$y_t = \text{ST} + \text{ART}_t + \text{S}_t, \ (t = 1, 2, 3, \ldots, j) \qquad [4.1]$$

It also makes the hypothesis of predictive relationships between latent ART factors, expressing an autoregressive function:

$$\text{ART}_t = \beta_{t-1}\,\text{ART}_{t-1} + \zeta_t, \ (\text{for } t = 2, 3, \ldots, j) \qquad [4.2]$$

where ζ_t represents the residual of the autoregressive effect.

Let us recapitulate: this model suggests articulating three latent factors (symbolized in Figure 4.1 by three circles from which three arrows point to each observed variable) likely to capture score variance on repeated measure of construct: (1) a "stable trait " factor (ST) capturing individual differences that are very durable over time, (2) "slow-changing traits" factors, called "autoregressive trait" (ART), capturing variance that demonstrates relatively gradual changes, ordered over time (a stability coefficient is provided from these factors), (3) finally, "state" (S) factors that capture variance unique and specific to each assessment occasion. These "state" factors reflect ephemeral, individual differences as well as measurement errors.

The objective of the STARTS model is to evaluate the part of each of these factors in construct score variance to determine its real nature. It goes without saying that if the construct is designed as a state-like (for example, an anxiety state), the part of variance of its scores imputable to the state (S) factor should be very substantial. However, if the construct is designed as trait-like (for example, anxiety trait), the part of variance of its scores attributable to the stable trait (ST) and to autoregressive trait should in be very substantial; also, autoregressive trait should display a good consistency over time, generally resulting in fairly high stability coefficients (b in Figure 4.1).

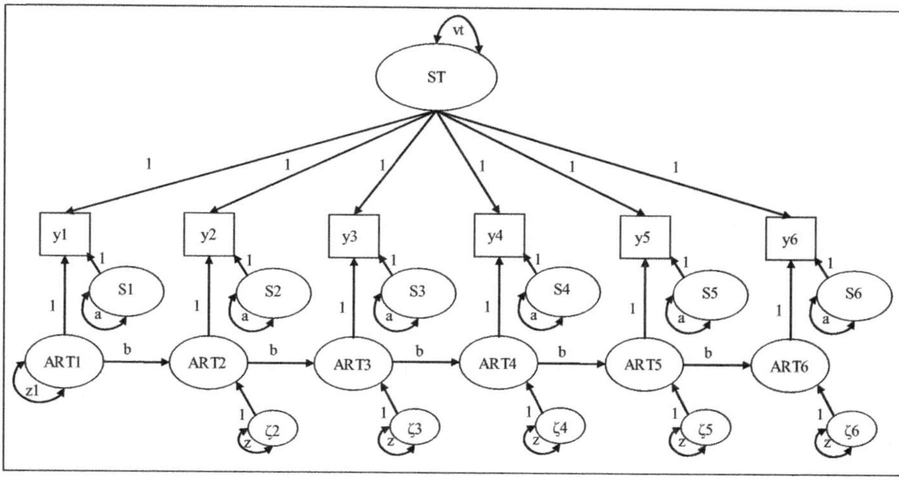

Figure 4.1. *Univariate STARTS model (the curved double arrows represent the variances)*

4.2.1.1.1. Identification of the STARTS model

To guarantee the model's mathematical identification, Kenny and Zautra [KEN 95] recommend fixing all the factor loadings at 1.00, constraining the autoregressive path coefficients to equality across time (b1 = b2..represented in Figure 4.1 by the same letter ("b"), constraining the ART factor residuals' variances (ζ) to equality across time (equality is represented by the same letter "z", allocated to each residual variable ζ in figure 4.1), and constraining state (S) factor variances to equality across time (equality by the same letter "a" in figure 4.1).

The simple autoregressive path reflects the fact that a variable's value over time t is the sum of the values of its proportion in t_{-1} and of a random residual (ζ). The equality between autoregressive paths makes it possible to guarantee stationarity, a condition sometimes necessary for analyzing the time series. A time series is called "stationary" (in a weak sense) if its statistical properties do not vary over time (expected value, variance, autocorrelation) [GRE 82]. However, the equality constraint between autoregressive paths is not always pertinent conceptually, and can therefore be lifted. It is usual to compare the two models, the one with the quality constraint with the one from which this constraint has been removed, and to retain the one that adjusts the data best.

Thus, independent of the number of assessment occasions, there are only five parameters to estimate in a univariate STARTS: four variances (those of ST, ART1, S, ζ; represented in Figure 4.1 by the curved double arrows: that is "vt", "z1", "a" and "z") and a stationary regression coefficient (β) (b). However, [KEN 01] suggest some constraints, especially those affecting the ART factors residuals' variances (ζ).

It will also be noted that there are, apart from the residual variables (S and ζ), which are also exogenous, two latent exogenous variables (ST and ART1), all recognizable by the fact they have no arrow pointing towards them. It should be remembered that in structural equation modeling, all the exogenous variables have a variance to be estimated or modeled (fix, constraint).

4.2.1.1.2. Specification of the STARTS model parameters in lavaan syntax

The model represented by Figure 4.1 has an observed, measured variable for the same participants on six occasions (y1, y2, y3, y4, y5 and y6), thus serving as indicators for the stable trait (ST). Section #1 of the specification illustrates this. The factorial loading of each of these variables has been fixed at 1.00, resulting in the hypothesis that a latent variable captures, in the same way, the variability present on each assessment occasion. In other words, the contribution of the "stable trait" factor is identical and constant from one assessment occasion to another. Thus, in lavaan syntax, a parameter fixed beforehand takes the value that precedes the asterisk placed in front of the parameter (for example 1*y1 = parameter fixed at 1.00, 5*y1 = parameter fixed at 0.50). Section #2 specifies the name given (vt*) to the variance of the latent variable ST (see Figure 4.1). This name will serve further to calculate the part of this factor's variance. Section #3 specifies the "autoregressive trait" factors linked to each assessment occasion. Loadings on each of these factors are also fixed at 1.00. Section #4 specifies the three autoregressive paths (regression coefficients, β) which are constrained to be equal over time (ART1 → ART2 = ART2 → ART3 = ART3 → ART4 = ART4 → ART5 = ART5 → ART6). Assigning a same letter (here "b") with an asterisk in front of parameters means labeling these parameters and at the same time constraining them to equality. Section #5 commands the constraint to the equality of all the variances of the latent ART variables' residuals (ζ), except ART1 which has no residual variable, (var(ζ_2) = var(ζ_3) = var(ζ_4) = var(ζ_5) = var(ζ_6)). We recall here that ζ is considered an exogenous variable whose variance must be estimated. It is the same for TRAIT, ART1 and S which are also exogenous variables (there is no arrow pointing directly towards them). It will be noted here that for each of the variances of ζ, the variables of which they are the residual must be used. For example, the var(ζ_2) is obtained by specifying an autocorrelation of the variable ART2 with itself (ART2 ~~ ART2), the var(ζ_3) = ART3 ~~ ART3. To constrain these variances to equality across/over time, it is enough to assign them a single letter, followed by an asterisk (here z*). It will be noted that as an exogenous variable, ART1 has a variance that is left free to be estimated (z1). Section #6 specifies the equality constraint of latent "state" (S) variable variances. For example, the command line (y1 ~~ a*y1) specifies the var(s1) as a free parameter to be estimated. The letter and the asterisk (a*) assigned to each variance name the parameter and announce its equality with those marked with the same letter (or name). Section #7 specifies the deletion of the correlation between the two latent exogenous variables ST and ART1, which, otherwise, will be correlated by default in lavaan.

STEP 2. Specification of the STARTS model with six assessment occasions (Figure 4.1).

```
model.SPE <- '
```

#1. Creating the latent "stable trait " factor (loadings fixed at 1.00).

```
ST =~ 1*y1 + 1*y2 + 1*y3 + 1*y4 + 1*y5 + 1*y6
```

#2. Naming the "stable trait " factor's variance (vt) for variance decomposition.

```
ST ~~ vt*ST
```

#3. Creating the latent "autoregressive trait" factors (loadings fixed at 1.00).

```
ART1 =~ 1*y1
ART2 =~ 1*y2
ART3 =~ 1*y3
ART4 =~ 1*y4
ART5 =~1*y5
ART6 =~1*y6
```

#4. Autoregressive paths constrained to equality (b*).

```
ART2 ~ b*ART1
ART3 ~ b*ART2
ART4 ~ b*ART3
ART5 ~ b*ART4
ART6 ~ b*ART5
```

#5. ζ variances constrained to equality (z*, except ART1, z1).

```
ART1 ~~ z1*ART1
ART2 ~~ z*ART2
ART3 ~~ z*ART3
```

```
ART4 ~~ z*ART4
ART5 ~~ z*ART5
ART6 ~~ z*ART6

#6. Latent "state" (S) variable variances constrained to equality (a*).

y1 ~~ a*y1
y2 ~~ a*y2
y3 ~~ a*y3
y4 ~~ a*y4
y5 ~~ a*y5
y6 ~~ a*y6

#7. Removal the default correlation between the two exogenous variables (0*).

ST ~~ 0*ART1

#8. Constraints.

z == z1 - (z1*b*b)
a > 0.001
z1 > 0.001
vt > 0.001

#9. Total variance decomposition of scores at "y".

var_total := vt + z1 + a
var_trait := vt/var_total
var_state := a/var_total
var_art := z1/var_total'
```

Section #8 specifies a constraint suggested by [KEN 01]. Although it is optional, this constraint affecting the residual variable variances (ζ) is included to ensure that the total variance of the "autoregressive trait" component (ART) is stable on all the assessment occasions. It is a stationarity constraint that may not only be necessary for identifying the model, but also convey the theoretical hypothesis that the distribution of the part of the variance imputable to this component remains constant and coherent over assessment occasions (i.e. stationarity assumption). It will be noted that "z1" is the name given to the variance of the latent variable ART1, "z" that given to the residual variables ζ variances, and "b" reflects the autoregressive path coefficients. The last three commands of section #8 specify the (optional) constraints whose objective is to prevent negative variances from appearing (i.e. forcing the variances to remain positive).

Section #9 specifies the way of calculating the total variance of the scores obtained from the variable measured ("y") and to estimate their component parts ("vt" = the name given to the variance of the latent variable "stable trait " (ST), "a" = the name given to the residual variables of the latent variables S; "var_trait" = variance of the latent trait factor (TS), "var_state" = variance of the latent state factor and of the measurement error (S), "var_art" = variance of the autoregressive trait factor. All these names ("vt", "z", "a", "b" etc.) are arbitrarily chosen by the researcher.

The simplicity of the univariate STARTS model however is marred by difficulties with convergence, which occur very often. This model is very useful where there is data at hand do not permit multivariate modeling. However, according to its designers [KEN 01] it needs several waves of assessment (10), a fairly large sample and additional constraints to remove the risk of improper solutions, to which it often falls victim. Not only are these additionnel constraints often complicated to specify, but they call above all for reservations and questions that are very much justified.

4.2.1.1.3. Decomposition of the total variance in measured variable (Y)'s scores

Total variance in each observed measure in time t is function of three sources : (1) stable trait (ST), (2) autoregressive trait (ART), and (3) ephemeral time-specific state and measurement error (S).

To estimate the part of each source in the total variance in the observed measure, it is enough to devide its specific variance estimate by total variance, which is the sum of the three variances (var(ST) + var (ART) + var (S_t)). Because of the equality constraints imposed on the model, such an operation is straightforward, even manually. We will demonstrate it using the following illustration.

4.2.1.1.4. Illustration

Specification of the STARTS model parameters

The STARTS model has been applied to the scores obtained at one dimension of the depression scale, that is CES-D, obtained on six occasions over a period of 13 years with a large sample. The total score for the items in this dimension (i.e. Depressed Affect, DA) is the variable measured on six occasions, CSD0DA, CSD3DA, CSD5DA, CSD8DA, CSD10DA, CSD13DA, each serving as an indicator for three factors: stable trait, autoregressive trait and state. The same constraints detailed previously have been applied to this model (see the box below).

A "robust" estimator (MLR) has been kept here to estimate the STARTS model. The missing data have been handled using the Full Information Maximum Likelihood (FIML) method implemented in lavaan via the command (missing = "fiml") and which has, among others, the advantage of providing unbiased parameter estimates and standard errors (on this subject, see [END 11, GRA 09]).

STEP 2.– Specifying the STARTS model (Figure 4.1) with six assessment occasions on one dimension of the CES-D scale.

```
model.SPE <- '
```

#1. Latent "stable trait " factor (loadings fixed at 1.00).

```
TRAIT =~ 1*CSD0DA + 1*CSD3DA + 1*CSD5DA + 1*CSD8DA +
1*CSD10DA + 1*CSD13DA
```

#2. Name (vt*) the variance of the stable trait for calculating the variance decomposition.

```
TRAIT ~~ vt*TRAIT
```

#3. Latent "autoregressive trait" factors (loadings fixed at 1.00).

```
ART1 =~ 1*CSD0DA
ART2 =~ 1*CSD3DA
ART3 =~ 1*CSD5DA
ART4 =~ 1*CSD8DA
ART5 =~ 1*CSD10DA
ART6 =~ 1*CSD13DA
```

#4. Autoregressive paths constrained to equality (b*).

```
ART2 ~ b*ART1
ART3 ~ b*ART2
ART4 ~ b*ART3
ART5 ~ b*ART4
ART6 ~ b*ART5
```

#5. ζ variances constraints to equality (z*, except ART1, z1).

```
ART1 ~~ z1*ART1
ART2 ~~ z*ART2
ART3 ~~ z*ART3
ART4 ~~ z*ART4
ART5 ~~ z*ART5
ART6 ~~ z*ART6
```

#6. Variances "state" (S) factors constrained to equality(a*).

```
CSD0DA ~~ a*CSD0DA
CSD3DA ~~ a*CSD3DA
CSD5DA ~~ a*CSD5DA
CSD8DA ~~ a*CSD8DA
CSD10DA ~~ a*CSD8DA
CSD13DA ~~ a*CSD13DA
```

#7. Remove the correlation by default between the two exogenous variables (*0).

```
TRAIT ~~ 0*ART1
```

#8. Constraints.

```
 z == z1 - (z1*b*b)
```

#9. Optional constraints to force the variances to be positive.

```
a > 0.001
z1 > 0.001
vt > 0.001
```

#10. Total variance decomposition in the measure scores.

```
var_total:= vt + z1 + a
var_trait:= vt/var_total
var_state:= a/var_total
var_art:= z1/var_total'
```

Evaluation of a univariate STARTS model solution

STEP 3. Estimating the STARTS model.

```
model.EST <- sem (model.SPE, data = BASE, missing =
"fiml", estimator = "MLR")
```

STEP 4. Obtaining the solution of the STARTS model.

```
summary (model.EST, fit.measures = T, std = T)
```

It will also be noted that for model estimation (Step 3) the "sem" model-fitting function has been sought.

We recall first that the variances have been forced to be positive (see section #9 of the model specification) with the aim of making the solution converge properly. It will be noted when examining Table 4.4 that the number of observations (participants) used (3694) was lower than the total of observations counted in our file (3777). The reason for this is as follows: lavaan automatically eliminates all the observations for which data are missing on all the variables used in the model. From this, it can be deduced that 83 participants had no score on all the variables in the model (CSD0DA, CSD3DA, CSD5DA, CSD8DA, CSD10DA, CSD13DA).

Overall goodness-of-fit indices

Evidently, the our sample size is not good news for χ^2, which has proven to be statistically significant. However, the other goodness-of-fit indices are clearly in favor of the univariate STARTS model: robust-CFI = 0.983, robust-TLI = 0.985 and robust-RMSEA = 0.022 (Table 4.4a).

Local fit indices

The section of the results that should draw our attention is the part showing "regressions" relating to autoregressive paths (Table 4.4b). The high path coefficients (0.796) indicate a high stability in time for the values of the variable measured. It will be noted that all the constraints imposed on the model lead to the total standardization of its coefficients (estimate = std.all), simplifying their interpretation by the same.

```
lavaan (0.5-23.1097) converged normally after 386 iterations

                                                  Used       Total
  Number of observations                          3694        3777

  Number of missing patterns                        49

  Estimator                                         ML      Robust
  Minimum Function Test Statistic               94.728      48.029
  Degrees of freedom                                17          17
  P-value (Chi-square)                           0.000       0.000
  Scaling correction factor                                  1.972
    for the Yuan-Bentler correction

Model test baseline model:

  Minimum Function Test Statistic             3491.920    1835.419
  Degrees of freedom                                15          15
  P-value                                        0.000       0.000

User model versus baseline model:

  Comparative Fit Index (CFI)                    0.978       0.983
  Tucker-Lewis Index (TLI)                       0.980       0.985

  Robust Comparative Fit Index (CFI)                         0.982
  Robust Tucker-Lewis Index (TLI)                            0.984

Loglikelihood and Information Criteria:

  Loglikelihood user model (H0)             -29135.833  -29135.833
  Scaling correction factor                                  0.623
    for the MLR correction
  Loglikelihood unrestricted model (H1)     -29088.468  -29088.468
  Scaling correction factor                                  1.796
    for the MLR correction

  Number of free parameters                         10          10
  Akaike (AIC)                               58291.665   58291.665
  Bayesian (BIC)                             58353.810   58353.810
  Sample-size adjusted Bayesian (BIC)        58322.035   58322.035

Root Mean Square Error of Approximation:

  RMSEA                                          0.035       0.022
  90 Percent Confidence Interval          0.028  0.042       0.017  0.028
  P-value RMSEA <= 0.05                          1.000       1.000

  Robust RMSEA                                               0.031
  90 Percent Confidence Interval                             0.021  0.042

Standardized Root Mean Square Residual:

  SRMR                                           0.061       0.061
```

Table 4.4a. *Overal goodness-of-fit indices of the univariate STARTS model*

Regressions:		Estimate	Std.Err	z-value	P(>|z|)	Std.lv	Std.all
ART2 ~							
ART1	(b)	0.796	0.162	4.917	0.000	0.796	0.796
ART3 ~							
ART2	(b)	0.796	0.162	4.917	0.000	0.796	0.796
ART4 ~							
ART3	(b)	0.796	0.162	4.917	0.000	0.796	0.796
ART5 ~							
ART4	(b)	0.796	0.162	4.917	0.000	0.796	0.796
ART6 ~							
ART5	(b)	0.796	0.162	4.917	0.000	0.796	0.796

Table 4.4b. *Autoregressive path estimates from the univariate STATRS (series) (contd.)*

Decomposition of the total variance of scores of the measured variable

Two rubrics (redudant, it is true) in the results enable us to make this decomposition. The first, that displaying the "Variances" (Table 4.4c), enables manual calculation of the parts of the proportion of variance of each of the three components of the STARTS in the total variance in the construct scores (TRAIT, ART and S). The total variance = vt + z1 + ve, which is 4.491 + 4.467 + 4.850 = 13.808.

The proportion of variance in scores in our measure at time *t* imputable to stable trait factor (TRAIT) is equal to 4.491/13.808 = 0.325, thus accounting for a rounded 33% of the total variance in this model. The proportion of variance imputable to autoregressive trait (z1) is equal to 4.467/13.808 = 0.324, which represent 32% of the total variance in this model. Remember that the autoregressive coefficients were high (0.796), indicating that even this variable part of depression leads to considerable stability in the short term. Finally, the proportion of variance imputable to the ephemeral state and to random error (ve) is equal to 4.850/13.808 = 0.351, which represent 35% of the total variance in this model. Finally, it will be noted that the TRAIT factor variance is slightly outside a 5% statistical significance bound ($p = 0.056$).

Variances:		Estimate	Std.Err	z-value	P(>|z|)	Std.lv	Std.all
TRAIT	(vt)	4.491	2.354	1.907	0.056	1.000	1.000
ART1	(z1)	4.467	2.069	2.159	0.031	1.000	1.000
.ART2	(z)	1.634	0.484	3.374	0.001	0.366	0.366
.ART3	(z)	1.634	0.484	3.374	0.001	0.366	0.366
.ART4	(z)	1.634	0.484	3.374	0.001	0.366	0.366
.ART5	(z)	1.634	0.484	3.374	0.001	0.366	0.366
.ART6	(z)	1.634	0.484	3.374	0.001	0.366	0.366
.CSD0DA	(ve)	4.850	0.417	11.641	0.000	4.850	0.351
.CSD3DA	(ve)	4.850	0.417	11.641	0.000	4.850	0.351
.CSD5DA	(ve)	4.850	0.417	11.641	0.000	4.850	0.351
.CSD8DA	(ve)	4.850	0.417	11.641	0.000	4.850	0.351
.CSD10DA	(ve)	4.850	0.417	11.641	0.000	4.850	0.351
.CSD13DA	(ve)	4.850	0.417	11.641	0.000	4.850	0.351

Table 4.4c. *Univariate STARTS variances (series) (contd.)*

The second rubric "Defined parameters" (Table 4.4d), obtained using the variance decomposition specification (section #10), automates the previous manual calculations.

```
Defined Parameters:
                   Estimate  Std.Err  z-value  P(>|z|)   Std.lv  Std.all
    var_total        13.808    0.380   36.305    0.000    6.850    2.351
    var_trait         0.325    0.170    1.910    0.056    0.146    0.425
    var_state         0.351    0.030   11.715    0.000    0.708    0.149
    var_art           0.324    0.149    2.167    0.030    0.146    0.425
```

Table 4.4d. *Variance decomposition of the univariate STARTS (series) (contd.)*

Reading Table 4.4d, it will be noted that the stable trait variance (var_trait) is significant is marginally significant at $p = 0.056$. This perhaps conveys the difficulties encountered by this model to lead to a "forceps" convergence after 386 iterations.

In total, the results of the STARTS model applied to data measuring a sub-dimension of depression over a period of 13 years show that two-thirds of the variance in scores obtained at this measure can be attributed to the trait, and a third is imputable to the state and to measurement error. Scores in this measure capture more depression-trait than depression-state.

As a conclusion, it seems clear to us that univariate STARTS is a heuristic model, conceptually imaginative, but empirically problematic. In fact, our sample size, the estimation of the model failed when using only four or five measurement times. In addition, the introduction of some constraints recommended by [KEN 01] prevented the models' convergence. Does this have something to do with its univariate nature based on only one observed error-free measure at each measurement occasion? This is what we propose to find out next by exploring its multivariate version.

4.2.1.2. *Multivariate STARTS*

By "multivariate", we should understand here the use of repeated measures of a construct that has multiple indicators (i.e. repeated measurement models). In other words, in place of modeling the total score on a measure, we model all the items either individually or in parcels, which can be subscores on construct's different dimensions. Figure 4.2 offers a diagram illustration of the multivariate STARTS. If we examine Figure 4.1 representing the univariate STARTS carefully, and compare it to Figure 4.2, it will be noted that the only difference between them lies in the substitution of manifest variables (y1, y2, y3, y4, y5, y6) by latent variables (DEP1, DEP2, DEP3, DEP4, DEP5, DEP6) each reflected by three indicators/items (the number of indicators depends on the measure used and the researcher's choices to aggregate or the items of this measure). These factors represent the construct on different occasions specific to the moment when the assessment is made. Each time, the same measurement model (i.e. CFA) underlyning the construct is present. It is therefore an extension of the univariate STARTS that, by integrating measurement models, offers the advantage of considering the reliability of scores in the observed measure of the construct [DON 12].

The variance decomposition of the latent variables representing the construct on each assessment occasion is the multivariate STARTS' main objective. The model makes the hypothesis that at each assessment point, the score variance depends on three concomitant factors (Figure 4.2): (1) a stable and enduring trait (TRAIT), (2) autoregressive trait (ART) whose change is slow, and (3) "STATE" factor. Given that measurement errors are considered in the measurement model, the "STATE" factor reflects the systematic variance of scores which is not imputable to the stable trait or to the autoregressive trait. The "STATE" variance thus becomes a real variance, specific to each assessment occasion and with measurement errors removed. Clearing the "STATE" factor of measurement errors is the main advantage of multivariate STARTS compared to univariate STARTS. As shown in Figure 4.2, the measurement errors linked to each indicator are correlated with one another over time to consider the unique shared variance over time. The measurement error variance contains both the measurement error variance and the variance unique (specific) to the indicator, which is not common to the latent variable's other indicators. It is this specific variance that we seek to control via autocorrelations of the same indicator's measurement errors over time. If this autocorrelation, which is due to factors other than the latent factor (the construct) on which the indicator depends, is not considered in the model, estimation of the construct's stability will be biased.

The variance decomposition of the latent variables representing the construct (DEP) on each assessment occasion invites us to pay particular attention to parameters in a multivariate STARTS model: (1) stable "TRAIT" variance ("vt" in the diagram in Figure 4.2) which indicates the proportion of variance of the scores repeated in the construct imputable to the stable and durable trait; (2) autoregressive trait factor variance (ART) which expresses these same scores' proportion of variance, which can be attributed to relatively stable and slowly changing factors; (3) "STATE" factor variance which conveys the proportion of variance of scores imputable to the specific occasion at the moment the assessment is taken; and finally (4) the autoregressive coefficients (beta) that express the temporal stability of autoregressive trait.

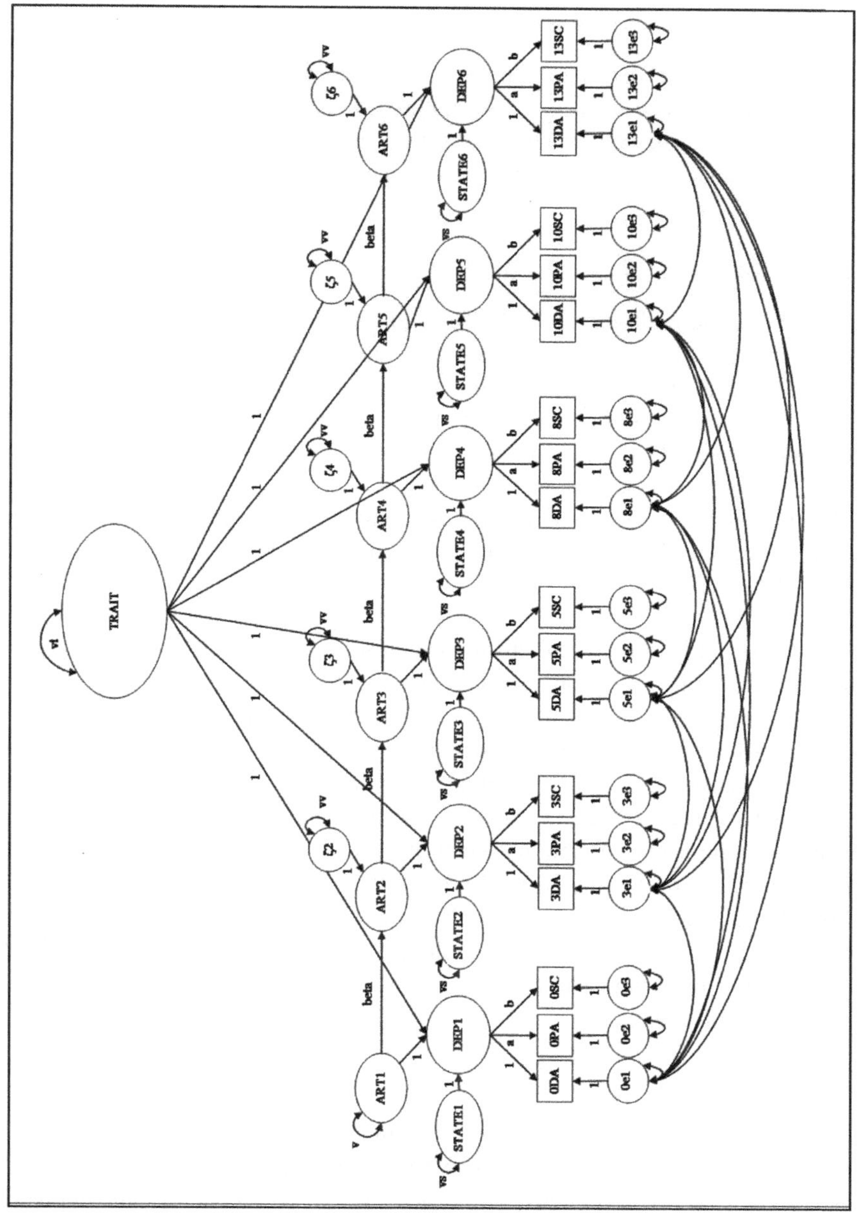

Figure 4.2. *Multivariate STARTS (the doubles curved arrows represent the variances and covariances). For clarity in the diagram, only the covariances between the measurement errors of the first indicator (DA) of the latent variables (DEP) have been drawn*

4.2.1.2.1. Decomposition of the total variance of the latent variable "occasion"

To estimate the proportion of each of the three initial parameters in the latent variable's total variance on each occasion, it is sufficient to divide its non-standardized value by the total variance which is the sum of three variances ($V_t + V_{ART} + V_{state}$). Thus, the variance of the stable trait is obtained as follows: $V_t = V_t/(V_t + V_{ART} + V_{state})$.

4.2.1.2.2. Identifying a multivariate STARTS

Identifying the multivariate STARTS requires several constraints to be imposed: (1) all STARTS' factors are independent of one another (orthogonal); (2) the trait (TRAIT) factor loadings are all fixed at 1.00; (3) the autoregressive trait factor (ART) loadings are fixed at 1.00; (4) the "STATE" factor loadings are fixed at 1.00; (5) the autoregressive trait factor variances(ζ) are constrained to equality in time; (6) the "state" factor variances are obliged to equality from one occasion; (7) the autoregressive (bêta) coefficients are constrained to equivalence through time; (8) other constraints (i.e. to be fixed initially) can impact the autoregressive trait factor variance. We will discuss this later.

In addition, at least one weak invariance must be imposed on the latent variable's indicators across the evaluation occasions (i.e. the loading of item 2 in *t1* = loading of item 2 in *t2* = loading of item 2 in *t3* = loading of item 2 in *t4*, etc.). Such an invariance, which means that the measure classes participants in the same way on each assessment occasion, is a minimum prerequisite for any longitudinal analysis. In fact, the absence of even a weak invariance means that the relationship between the indicator and the latent variable on which it depends changes with time, suggesting that this indicator potentially measures a construct that differs at each assessment. Finally, it is also advisable to autocorrelate the same indicator's measurement errors through time to take account of the shared method variance (for example, the measurement error of item 1 in *t1* correlated with the measurement error of item 1 in *t2*, with item 1 in *t3*, etc.).

4.2.1.2.3. Illustration

Here, we return to the data that served to illustrate the univariate STARTS model that we convert here into a multivariate STARTS. We will use the total scores at each of the CES-D scale dimensions as indicators of the latent variable depression (DEP) on six assessment occasions (Figure 4.2). For example, depression in t_1 (DEP1) has as indicators (measured variables), scores for the "Depressed Affect, DA" dimension (0DA), scores for the "Positive Affect, PA" (0PA) dimension and scores for the "Somatic Complaints, SC" (0SC) dimension for the CES-D scale.

Specifying parameters for the multivariate STARTS model in lavaan syntax

For didactic reasons, we have also chosen here to partition the specification of our multivariate STARTS into numbered sections, the number of one section is joined to a hash sign (#1, #2, #3 etc.). Presented in the box below, we will review them each in turn.

STEP 2. Specifying the multivariate STARTS model with six time measurements of the CES-D scale (Figure 4.2).

```
model.SPE <- '
```

#1. Models for Measurement models of depression at each time with weak invariance (i.e. factor laodings' equality, as indicated in Figure 4.2).

```
DEP1 =~ 1*CSD0DA + a*CSD0PA + b*CSD0SC
DEP2 =~ 1*CSD3DA + a*CSD3PA + b*CSD3SC
DEP3 =~ 1*CSD5DA + a*CSD5PA + b*CSD5SC
DEP4 =~ 1*CSD8DA + a*CSD8PA + b*CSD8SC
DEP5 =~ 1*CSD10DA + a*CSD10PA + b*CSD10SC
DEP6 =~ 1*CSD13DA + a*CSD13PA + b*CSD13SC
```

#2. Factor "stable trait" (loadings fixed at 1.00).

```
TRAIT =~ 1*DEP1 + 1*DEP2 + 1*DEP3 + 1*DEP4 + 1* DEP5 +
1*DEP6
```

#3. Name the trait factor variance ("vt" for example) for the variance decomposition.

```
TRAIT ~~ vt*TRAIT
```

#4. Autoregressive trait factors (loadings fixed at 1.00).

```
ART1 =~ 1*DEP1
ART2 =~ 1*DEP2
ART3 =~ 1*DEP3
ART4 =~ 1*DEP4
ART5 =~ 1*DEP5
ART6 =~ 1*DEP6
```

#5. "State" factors (loadings fixed at 1.00).

```
STATE1 =~ 1*DEP1
STATE2 =~ 1*DEP2
STATE3 =~ 1*DEP3
STATE4 =~ 1*DEP4
STATE5 =~ 1*DEP5
STATE6 =~ 1*DEP6
```

#6. Autoregressive paths (arbitrarily called "beta" here) constrained to equality (*via* beta*).

```
ART2 ~ beta*ART1
ART3 ~ beta*ART2
ART4 ~ beta*ART3
ART5 ~ beta*ART4
ART6 ~ beta*ART5
```

#7. The latent DEP variable variances fixed at zero (as each is provided with a residual/distrubance variable by default).

```
DEP1 ~~ 0*DEP1
DEP2 ~~ 0*DEP2
DEP3 ~~ 0*DEP3
DEP4 ~~ 0*DEP4
DEP5 ~~ 0*DEP5
DEP6 ~~ 0*DEP6
```

#8. Latent "STATE" variable variances constrained to equality.

```
STATE1 ~~ vs*STATE1
STATE2 ~~ vs*STATE2
STATE3 ~~ vs*STATE3
STATE4 ~~ vs*STATE4
STATE5 ~~ vs*STATE5
STATE6 ~~ vs*STATE6
```

#9. Latent "ART" variable variances constrained to equality (except those of ART1).

```
ART1 ~~ v*ART1
ART2 ~~ vv*ART2
ART3 ~~ vv*ART3
```

```
ART4  ~~  vv*ART4
ART5  ~~  vv*ART5
ART6  ~~  vv*ART6
```

#10. Remove the correlations by default between the model's exogenous variables (*0).

```
TRAIT  ~~  0*ART1
TRAIT  ~~  0*STATE1
TRAIT  ~~  0*STATE2
TRAIT  ~~  0*STATE3
TRAIT  ~~  0*STATE4
TRAIT  ~~  0*STATE5
TRAIT  ~~  0*STATE6
ART1  ~~  0*STATE1
ART1  ~~  0*STATE2
ART1  ~~  0*STATE3
ART1  ~~  0*STATE4
ART1  ~~  0*STATE5
ART1  ~~  0*STATE6
STATE1  ~~  0*STATE2
STATE1  ~~  0*STATE3
STATE1  ~~  0*STATE4
STATE1  ~~  0*STATE5
STATE1  ~~  0*STATE6
STATE2  ~~  0*STATE3
STATE2  ~~  0*STATE4
STATE2  ~~  0*STATE5
STATE2  ~~  0*STATE6
STATE3  ~~  0*STATE4
STATE3  ~~  0*STATE5
STATE3  ~~  0*STATE6
STATE4  ~~  0*STATE5
STATE4  ~~  0*STATE6
STATE5  ~~  0*STATE6
```

#11. Autocorrelations of measurement errors over time (45 correlations).

```
CSD0DA  ~~  CSD3DA + CSD5DA + CSD8DA + CSD10DA + CSD13DA
CSD3DA  ~~  CSD5DA + CSD8DA + CSD10DA + CSD13DA
CSD5DA  ~~  CSD8DA + CSD10DA + CSD13DA
CSD8DA  ~~  CSD10DA + CSD13DA
```

```
CSD10DA ~~ CSD13DA
CSD0PA ~~ CSD3PA + CSD5PA + CSD8PA + CSD10PA + CSD13PA
CSD3PA ~~ CSD5PA + CSD8PA + CSD10PA + CSD13PA
CSD5PA ~~ CSD8PA + CSD10PA + CSD13PA
CSD8PA ~~ CSD10PA + CSD13PA
CSD10PA ~~ CSD13PA
CSD0SC ~~ CSD3SC + CSD5SC + CSD8SC + CSD10SC + CSD13SC
CSD3SC ~~ CSD5SC + CSD8SC + CSD10SC + CSD13SC
CSD5SC ~~ CSD8SC + CSD10SC + CSD13SC
CSD8SC ~~ CSD10SC + CSD13SC
CSD10SC ~~ CSD13SC

#12. Constraints.

vv == v - (v*beta*beta)
vt > 0.001
vs > 0.001
vv > 0.001

#13. Decomposition of the total variance of the latent variable DEP.

var_total:= vt + vs + v
var_trait:= vt/var_total
var_state:= vs/var_total
var_ART := v/var_total'
```

The first section #1 specifies the six measurement occasions of the depression scale (DEP1... DEP6). On each occasion, depression has been measured by the same three indicators. A weak temporal invariance has been imposed on two indicators, since the first has been fixed at 1.00 to identify the model. For example, factor loading of CSD0PA = factor loading of CSD3PA = the loading of CSD5PA = the loading of CSD8PA = factor loading of CSD10PA = the loading CSD13PA on their respective latent variables. The letter "a*" allocated to each of the loadings indicates this equality constraint.

Section #2 introduces the latent "stable trait" factor, called "TRAIT" here. It weighs equally on each of the six latent DEP variables (loadings fixed at 1.00).

Section #3 serves to name the TRAIT factor variance. The name given to this factor (vt*) will be used to partition the total variance of the latent variable DEP.

Section #4 specifies the “autoregressive trait” (called “ART” here) factors. On each assessment occasion, the construct depression (DEP) undergoes the effect of an “autoregressive trait” factor (ART) whose loading is fixed at 1.00.

Section #5 specifies the latent “state” variables (called “STATE”). Each latent DEP variable undergoes the effect of a latent STATE variable whose loading is fixed at 1.00.

Section #6 specifies the autoregressive paths that are constrained to equality over time (the repeated presence of “beta*” indicates this equality constraint.

Section #7 fixes the variance of each of the latent DEP variables at zero. There are many reasons for this. First, as endogenous variables, lavaan will automatically and by default assign to each one a residual/disturbance variable. However, this has been replaced by our latent “STATE” variable (section #5). Thus, to avoid duplication, it is vital to require lavaan not to introduce residual variables assigned to latent endogenous DEP variables into the model by default. For example, “DEP1 ~~ 0*DEP1” indicates cancellation of the disturbance variable logically linked to the latent DEP1 variable.

Section #8 specifies the equality constraint on the variances of all the latent “STATE” variables. The sign “vs*”, which they share, conveys this equality requirement.

Section #9 specifies the equality between the latent ART variable variance. In reality, apart from ART1 whose variance is free, the other latent ART variables (ART2… ART6) do not strictly speaking have variance, as they are exogenous. Thus, it is a question of constraining the equality rather on the variances of their respective disturbance variables (ζ). The letter they share, “vv*”, expresses this equality constraint.

Section #10 is important, as it specifies the cancelation of correlations between the model’s exogenous variables, correlations programmed by default in lavaan. Our model contains three latent exogenous variables, which can be recognized as there is no arrow pointing towards them (Figure 4.2): the TRAIT variable, the ART1 variable and the STATE variables that have ceased to be residuals of the DEP variables to become true latent variables specified in our syntax (section #5). And as figures 4.1 and 4.2 show, the STARTS model assumes independence across the latent exogenous variables, hence the need to delete their intercorrelations: correlation between TRAIT and ART1, correlations between TRAIT and STATE1… STATE6, correlations between ART1 and STATE1… STATE6, and finally intercorrelations between the STATE variables.

Section #11 specifies the measurement error autocorrelations for each indicator over time. Precisely, there are 15 autocorrelations for each indicator measured six times (6*(6 – 1)/2 = 15). For instance, the measurement error of the CSD0DA indicator is correlated with each measurement error of this indicator measured on the following occasions: CSD3DA, CSD5DA, CSD8DA… CSD13DA, see Figure 4.2). It has already been seen that the measurement error encloses a sort of combination of random measurement error and specific indicator/item variance. And because this specific indicator variance is assumed to be enduring (not random), it could be appropriate to autocorrelate it over time. Since it is not random, but rather systematic, this measurement error may occur again at each measurement occasion. Such an error could reflect method effect. However, these autocorrelations are sometimes neither necessary nor justified.

Section #12 specifies constraints that are similar to those specified for the univariate STARTS.

Section #13 shows the way to calculate and partition the total variance of repeated measures. The total variance concerns the latent DEP variables, unlike the univariate STARTS for which the calculation of the total variance concerns the observed variable "y".

Model evaluation

STEP 3. Estimation of the multivariate STARTS model.

```
model.EST <- sem (model.SPE, data = BASE, missing =
"fiml", estimator = "MLR")
```

STEP 4. Retrieving the results of the multivariate STARTS model.

```
summary (model.EST, fit.measures = T, std = T)
```

A "robust" estimator (MLR) has been chosen to estimate the multivariate STARTS model. The missing data have been handled using the Full Information Maximum Likelihood (FIML).

Overall goodness-of-fit indices

Despite a statistically significant χ^2 (equal to 19542 with 102 degrees of freedom), the values of the other fit indices (TLI, CFI, RMSEA) indicate that the model is in harmony with the data (Table 4.5). This observation enables us to examine the local fit indices.

```
lavaan (0.5-23.1097) converged normally after 1164 iterations

                                                  Used       Total
  Number of observations                          3707        3777

  Number of missing patterns                       197

  Estimator                                         ML      Robust
  Minimum Function Test Statistic              273.461     195.462
  Degrees of freedom                               102         102
  P-value (Chi-square)                           0.000       0.000
  Scaling correction factor                                  1.399
    for the Yuan-Bentler correction

Model test baseline model:

  Minimum Function Test Statistic            19170.008   13515.928
  Degrees of freedom                               153         153
  P-value                                        0.000       0.000

User model versus baseline model:

  Comparative Fit Index (CFI)                    0.991       0.993
  Tucker-Lewis Index (TLI)                       0.986       0.990

  Robust Comparative Fit Index (CFI)                         0.993
  Robust Tucker-Lewis Index (TLI)                            0.990

Loglikelihood and Information Criteria:

  Loglikelihood user model (H0)             -78138.108  -78138.108
  Scaling correction factor                                  1.135
    for the MLR correction
  Loglikelihood unrestricted model (H1)     -78001.378  -78001.378
  Scaling correction factor                                  1.421
    for the MLR correction

  Number of free parameters                         87          87
  Akaike (AIC)                              156450.217  156450.217
  Bayesian (BIC)                            156991.181  156991.181
  Sample-size adjusted Bayesian (BIC)       156714.737  156714.737

Root Mean Square Error of Approximation:

  RMSEA                                          0.021       0.016
  90 Percent Confidence Interval          0.018  0.024       0.013  0.019
  P-value RMSEA <= 0.05                          1.000       1.000

  Robust RMSEA                                               0.019
  90 Percent Confidence Interval                             0.015  0.023

Standardized Root Mean Square Residual:

  SRMR                                           0.056       0.056
```

Table 4.5a. *Overall goodness-of-fit indices from the multivariate STARTS*

Local fit indices and variance decomposion

It can be seen from the results of the "Regressions" rubric (Table 4.5b) that the standardised coefficients of the autoregressive paths (0.711) indicate high stability over time of the latent variable DEP scores.

```
Regressions:
                      Estimate  Std.Err  z-value  P(>|z|)   Std.lv  Std.all
  ART2 ~
    ART1     (beta)      0.711    0.059   11.968    0.000    0.711    0.711
  ART3 ~
    ART2     (beta)      0.711    0.059   11.968    0.000    0.711    0.711
  ART4 ~
    ART3     (beta)      0.711    0.059   11.968    0.000    0.711    0.711
  ART5 ~
    ART4     (beta)      0.711    0.059   11.968    0.000    0.711    0.711
  ART6 ~
    ART5     (beta)      0.711    0.059   11.968    0.000    0.711    0.711
```

Table 4.5b. *Autoregessive path estimates of the multivariate STARTS (contd.)*

As for the scores' total variance decomposition, the results of the "Defined Parameters" rubric (Table 4.5c) we learn that the proportion of the "stable trait" factor is the highest (0.473 which is 47%). That of the autoregressive trait is 29%, whereas the proportion of the "state" factor represents 24% (0.239). All these parts of total variance are statistically significant to $p = 0.000$.

```
Defined Parameters:
                   Estimate  Std.Err  z-value  P(>|z|)   Std.lv  Std.all
    var_total         9.564    0.353   27.127    0.000    3.000    3.000
    var_trait         0.473    0.029   16.316    0.000    0.333    0.333
    var_state         0.239    0.029    8.230    0.000    0.333    0.333
    var_ART           0.288    0.037    7.866    0.000    0.333    0.333
```

Table 4.5c. *Multivariate STARTS variance decomposition (contd.)*

To conclude, we will say here that three quarters (76%) of the score variance for our depression indicators are imputable to the stable trait and to autoregressive trait (slow-changing component). Thus, the scores in this measure seem to capture depresion-trait than depression-state.

4.2.2. *The Trait-State-Occasion Model*

Suggested by Cole, Martin and Steiger [COL 05], the Trait-State-Occasion (TSO) model is intended to be a multivariate extension of the univariate STARTS. This extension suggested by these authors in 2005, well before that of the multivariate STARTS suggested by Donnellan and his colleagues [DON 12]. Figure 4.3 offers a visualization in diagram form of the TSO. It looks exactly like the multivariate STARTS. However, it is nothing of the sort. We will detail the TSO's constituent parts first, and then reveal its dissimilarities from the STARTS.

Each time assessment is taken (*t*), the TSO requires multiple indicators/items (at least two) to represent the construct to be studied across time. Each time assessment is taken, these same indicators serve to evaluate the state ("S_DEP" in Figure 4.3) in which the participant is on the construct measured. The latent variable ("S_DEP") therefore represents assessment of the construct at moment *t*. Thus, each indicator is subject to the effect of this latent variable underlying the construct and to the effect of measurement error, $Y_{it} = S_{it} + \delta_{it}$ (i.e. a CFA). This latent variable ("S_DEP") is, itself, affected at time *t*, by two influences: that of a trait ("T_DEP") factor supposed to be time-invariant factor, and that of time-varying factor, as it reflects the specific moment when the construct is measured (i.e a second-order CFA). This last factor, called "occasion" ("occas" in Figure 4.3), represents the specific circumstances at the moment the assessment is made and which affect its scores. Both these effects are reflected in the two arrows pointing to S_DEP (thus, S_DEP = T_DEP + occas). It will be noted here that the ("occas") factors are no other than the residual variables necessarily linked to the first order variables in a hierarchical model (Figure 4.3). In fact, each endogenous variable (which has at least one arrow directed towards it) is necessarily provided with a residual variable (measurement error, residual error, disturbance). For example, ζ_2, ζ_3, ζ_4, ζ_5, ζ_6 in Figure 4.3 represent the residual variables of $occas_2$, $occas_3$, $occas_4$, $occas_5$, $occas_6$ respectively. The $occas_t$ variables are related to one another by an autoregressive process through the effect of the first occasion on the subsequent one ($occas_t \rightarrow occas_{t+1}$). The autoregressive coefficients (β) can thus express the persistence in over time of the effects of specific circumstances related to the construct assessment context.

Thus, the TSO's constituent components make it possible to describe the construct's true nature (quasi-trait or quasi-state), to specify at what point the occasion-related circumstances are persistent over time, according to B in the autoregressive function, and to determine the unexplained proportion of variance, accounting for the two previous components (i.e., Trait and Occasion).

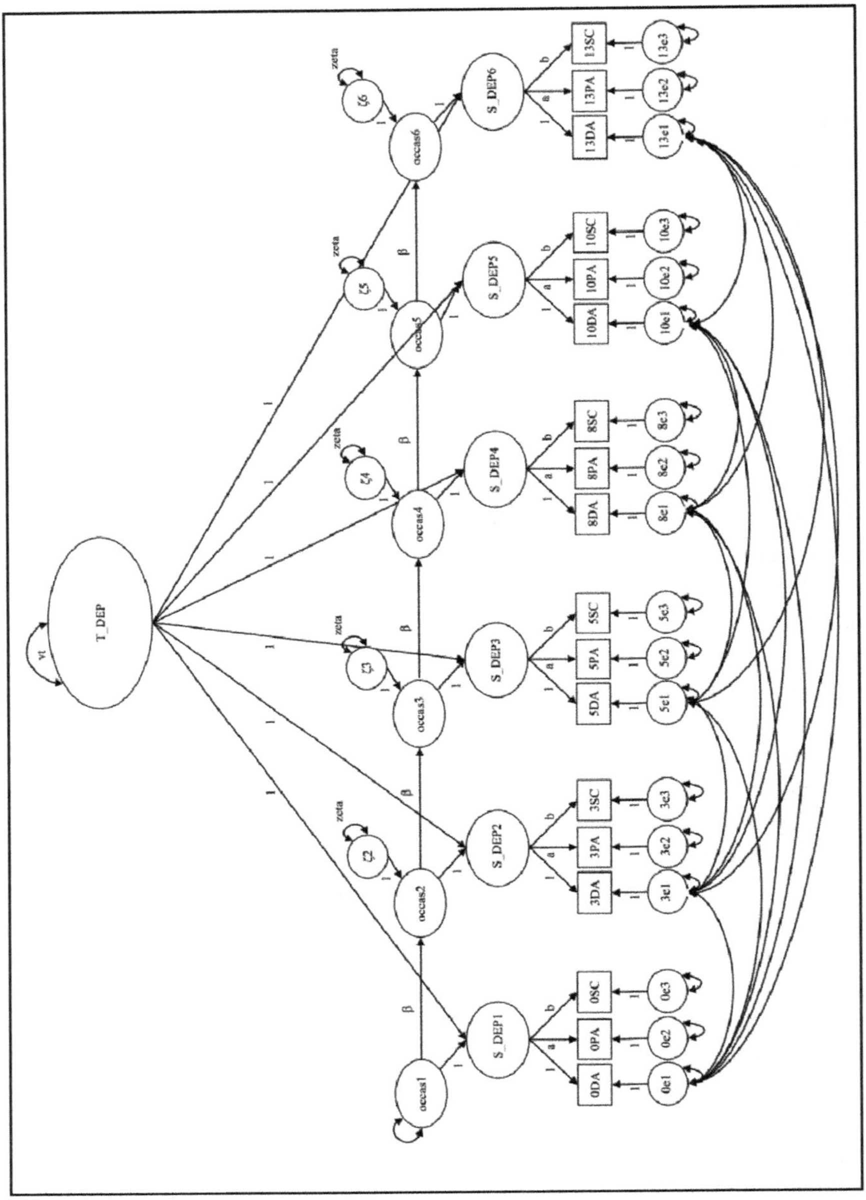

Figure 4.3. *Trait-state-occasion-TSO model (occas = occasion factor, S_DEP = state factor) (the curved double arrows represent the variances and covariances). For clarity on the diagram, only the covariances between the measurement errors of the first indicator (DA) of the latent variables (S-DEP) have been drawn*

The only difference between the TSO and the multivariate STARTS lies in the disappearance in the TSO of the latent "state" variable available in STARTS. The latent variable "occasion" in STARTS becomes the latent variable "state" in TSO. This is therefore without disturbance variable (without residual). The "occasions" variables are in a way phantom variables and not residual variables as such. The proof is that from $occas_2$, each is allocated a residual variable (ζ).

4.2.2.1. *Decomposition of the latent variable's total variance "state"*

It has already been pointed out that the construct measured each time *t* ("S-DEP" in Figure 4.3) is subject to two influences (represented by the two arrows in the diagram): the influence of the T_DEP trait, which is time-invariant factor, and the influence of occasion-related circumstances ($occas_t$), a time-varying factor. And because there is no disturbance term associated to S_DEP_t, both these effects share the total variance of scores (S_DEP_t) obtained at time *t*; the total variance (Var(TT)) = Var(T) + Var(O_t). The percentage of variance imputable to the stable trait = Var(T)/Var(TT), the remainder being the proportion of variance of the scores obtained at a moment *t* not explained by the trait, but imputable to the context (occasion) of assessment. However, it makes sense to specify that only the "stable trait" and the first "occasion" factor ($occas_1$) variance are estimated (recognizable in the figures by their curved double arrows) and allocated the p-values, as they are exogenous (independent) variables. As endogenous variables, the following "occasion" factors ($occas_2$, $occas_3$... $occas_{t+1}$) have no variance, but the residual variables (ζ) which are allocated the variance (the curved double arrows provide a visual illustration of it). The variances of "occasion" factors subsequent to the first factor ($occas_1$) depend on two sources of influence: the autoregressive effect (β) of the previous "occasion" factor ($occas_1 \rightarrow occas_2$) and the effect of the residual variable (ζ) linked to each endogenous "occasion" factor. The variances of $occas_2$, $occas_3$... $occas_6$ are not estimated automatically by SEM software programs, including lavaan. We will illustrate this statement further.

4.2.2.2. *Identification of the TSO*

Cole and his colleagues [COL 05] make the following recommendations and suppositions to eventually simplify the model and facilitate its convergence: (1) all the TSO's exogenous factors are independent of one another (orthogonal); (2) the factor loadings of the trait factor (T_DEP) are all fixed at 1.00; (3) the loadings of the "occasions" factors are fixed at 1.00; (4) the autoregressive coefficients (β) linking the "occasion" factors are constrained to equality across time (homogeneity of autoregressive effects); (5) the residual variable variance (ζ) of the "occasion" factors are constrained to equality across time (homogeneity of variances).

In addition, it is recommended to impose at least a weak measurement invariance across time (i.e. the loading of item 2 in *t*1 = loading of item 2 in *t*2 = loading of item 2 in *t*3 = loading of item 2 in *t*4). Such an invariance, which means that the on each assessment occasion, is a classes the participants in the same way at each assessment occasion, is taken, is a minimum prerequisite for any longitudinal analysis. Finally, it is also recommended to autocorrelate the measurement errors of the same indicator over time to account for the method's shared variance (for example, the measurement error of item 1 in *t*1 correlated with the measurement of error item 1 in *t*2, with item 1 in *t*3, etc.) Failure to control this method effect results in an an overestimation of temporal stability, which then runs the risk of generating overevaluated and biased estimations on trait factor variance and the stability of the "occasion" factors.

Because they are not actually necessary but merely recommended, some constraints, which aim to simplify the model and facilitate its convergence, can be relaxed when they prove to have no conceptual base. For example, it is sometimes conceptually more plausible not to assume the autoregressive effects (β) are homogenous (equal) or to assume that the residual variances are not constant over time. Also, it is permissible to suppose that the "stable trait" factor does not occur identically over time, and to remove the equality constraint that weighs on the loadings (all fixed at 1.00). The approach to adopt here consists of comparing models (those with constraints versus those without) and ensuring that the constraints do not worsen model fit quality. We can even compare the TSO with a second-order CFA model (i.e. remove the "occasion" factors and restore the residual variables of the first order latent variables).

4.2.2.3. *Illustration of a TSO*

We will apply the TSO to the data used to illustrate the previous STARTS models. Each of the six latent variables (S_DEP) in our 6-wave time series has three indicators measuring the depressive state at time *t* (DA, PA and SC, reflecting the scores for each of the three CES-D dimensions). A measurement error is associated to each indicator (e). We will now see how to convert the elements making up the diagram on Figure 4.3 in lavaan syntax.

4.2.2.3.1. Specification of TSO parameters in lavaan syntax

The specification of the TSO illustrated by the diagram on Figure 4.3 is broken down into eight sections.

STEP 2. Specification of the TSO model with six assessment occasions of the CES-D scale is measured (Figure 4.3).

```
model.SPE <-  '
```

#1. Latent “state-depression: S_DEP” variables.

```
S_DEP1 =~ 1*CSD0DA + a*CSD0PA + b*CSD0SC
S_DEP2 =~ 1*CSD3DA + a*CSD3PA + b*CSD3SC
S_DEP3 =~ 1*CSD5DA + a*CSD5PA + b*CSD5SC
S_DEP4 =~ 1*CSD8DA + a*CSD8PA + b*CSD8SC
S_DEP5 =~ 1*CSD10DA + a*CSD10PA + b*CSD10SC
S_DEP6 =~ 1*CSD13DA + a*CSD13PA + b*CSD13SC
```

#2. “trait-depression: T_DEP” factor (loadings fixed at 1.00).

```
T_DEP =~ 1*S_DEP1 + 1*S_DEP2 + 1*S_DEP3 + 1*S_DEP4 +
1*S_DEP5 + 1*S_DEP6
```

#3. removal of residual variables associated automatically to latent variables (S_DEP).

```
S_DEP1 ~~ 0*S_DEP1
S_DEP2 ~~ 0*S_DEP2
S_DEP3 ~~ 0*S_DEP3
S_DEP4 ~~ 0*S_DEP4
S_DEP5 ~~ 0*S_DEP5
S_DEP6 ~~ 0*S_DEP6
```

#4. Creation of latent “occasion: occas” variables.

```
occas1 =~ 1*S_DEP1
occas2 =~ 1*S_DEP2
occas3 =~ 1*S_DEP3
occas4 =~ 1*S_DEP4
occas5 =~ 1*S_DEP5
occas6 =~ 1*S_DEP6
```

#5. The autoregressive paths/effects of an “occasion” on the following occasion.

```
occas2 ~ bêta*occas1
occas3 ~ bêta*occas2
occas4 ~ bêta*occas3
occas5 ~ bêta*occas4
occas6 ~ bêta*occas5
```

#6. Removal the default correlations between the models’ latent exogenous variables (*0).

```
T_DEP ~~ 0*occas1

#7. Constrain the residual variable variance to equality (ζ; the first, is not residual, figure 4.3).

occas1 ~~ occas1
occas2 ~~ zeta*occas2
occas3 ~~ zeta*occas3
occas4 ~~ zeta*occas4
occas5 ~~ zeta*occas5
occas6 ~~ zeta*occas6

#8. Measurement error autocorrelations (45 correlations).

CSD0DA ~~ CSD3DA + CSD5DA + CSD8DA + CSD10DA + CSD13DA
CSD3DA ~~ CSD5DA + CSD8DA + CSD10DA + CSD13DA
CSD5DA ~~ CSD8DA + CSD10DA + CSD13DA
CSD8DA ~~ CSD10DA + CSD13DA
CSD10DA ~~ CSD13DA
CSD0PA ~~ CSD3PA + CSD5PA + CSD8PA + CSD10PA + CSD13PA
CSD3PA ~~ CSD5PA + CSD8PA + CSD10PA + CSD13PA
CSD5PA ~~ CSD8PA + CSD10PA + CSD13PA
CSD8PA ~~ CSD10PA + CSD13PA
CSD10PA ~~ CSD13PA
CSD0SC ~~ CSD3SC + CSD5SC + CSD8SC + CSD10SC + CSD13SC
CSD3SC ~~ CSD5SC + CSD8SC + CSD10SC + CSD13SC
CSD5SC ~~ CSD8SC + CSD10SC + CSD13SC
CSD8SC ~~ CSD10SC + CSD13SC
CSD10SC ~~ CSD13SC'
```

The first section #1 specifies the six latent variables (STATE) defined by the same indicators at each assessment occasion. A weak temporal invariance has been imposed on two indicators, as the first was fixed at 1.00 to identify the measurement model. These factors assess the level of the participant's depressive state at time *t*. Thus, "Var(S_DEP_1)" represents the variance of depression scores in t_1.

Section #2 specifies the single stable factor, depression-trait, called "T_DEP", which explains identically (i.e. so that the loadings are fixed at 1.00), each depression-state factor (S_DEP).

Section #3 specifies the deletion of residual variables associated automatically by lavaan to the endogenous variables. In effect, as soon as each of the latent variables S_DEP1 to S_DEP6 undergoes trait factor (T_DEP) effect, they become endogenous variables and consequently are automatically allocated the residual variables.

Section #4 specifies creation on the remains of previously erased residual variables and, to avoid duplication, the creation of six latent "occasion" variables defining one corresponding latent variable S_DEP ($occas_1 \rightarrow$ S_DEP1… $occas_6 \rightarrow$ S_DEP6) at assessment point. The "occasion" factor represents the proportion of variance in depression not explained by the "stable trait" (T_DEP) factor in time *t*.

Section #5 reveals the syntax of the autoregressive effects that are constrained to equality across time (homogeneity of autoregressive effects).

Section #6 specifies orthogonality across the TSO's latent exogenous variables. There are two: T_DEP and $occas_1$ for which we must remove the correlation, which is considered in the model by default. The other "occas" variables have become endogenous because of the autoregressive process.

Section #7 specifies the equality constraints on the residual variables (ζ_t) of the variance of the latent endogenous occasion variables (homogeneity of variances). Recall here that "occas1 ~~ occas1" specifies the latent $occas_1$ variable variance which, as an exogenous variable, does not have a residual variable. While for example, "occas2 ~~ zeta*occas2" specifies the variance of the residual variable (ζ_2) of the variable $occas_2$ which, as a latent endogenous variable has no variance to estimate. The autoregressive process has rendered variables $occas_2$ to $occas_6$ endogenous and so each does have a residual variable (ζ). These residual variables are automatically considered by lavaan, hence their absence from the model specification (and their presence in the diagram).

Section #8 specifies the measurement error autocorrelations for each repeated measure, as it precisely 15 autocorrelations for each indicator measured six times ($6*(6-1)/2 = 15$) (i.e. the CSD0DA indicator measurement error is correlated with the same indicator assessed on the assessment occasions: CSD3DA, CSD5DA, CSD8DA… CSD13DA; see Figure 4.3).

4.2.2.3.2. Model evaluation

STEP 3. Model estimation TSO.

```
model.EST <- sem (model.SPE, data = BASE, missing =
"fiml", estimator = "MLR")
```

STEP 4. Retrieving the results of the TSO.

```
summary (model.EST, fit.measures = T, std = T,
rsq = T)
```

A "robust" estimator (MLR) has been used to estimate the TSOmodel, and the missing data have been processed using the Full Information Maximum Likelihood (FIML) method.

Overall Goodness-of-fit indices

Notwithstanding a significant χ^2, as one might expect, the TSO fits the data very well, as the values of the robust goodness-of-fit indices bear witness: robust-CFI = 0.994, robust-TLI = 0.990, robust-RMSEA = 0.018.

Local fit indices

When we examine the parameters' estimates, they reveal no offending and inadmissible values, which enables us to conclude that the solution is proper (Table 4.6b). It will be revealed here that the autoregressive coefficients are statistically significant, but low in size (from 0.207 to 0.237). These coefficients reflect that context-specific influences on the measure are stable. We note here that a squared standardized autoregressive coefficient (β^2) expresses its proportion in the variance of the variable on which it acts. For example, the effect of $occas_1$ on $occas_2$ which is 0.237, means that the contribution of $occas_1$ in the variance of $occas_2$ is 5.6% $[(0.237)^2*100]$. Moreover, examination of the "R-Square" results rubric enables us to verify it. In fact, the R^2 of $occas_2$ is 0.056; and as $occas_1$ is the only influence affecting $occas_2$, it can be deduced that 5.6% (= 0.056*100) of its variance depends on this sole autoregressive effect, the remainder (94.4%) depends on the residual variance (ζ_2). The R^2 of $occas_3$ is 0.044. It corresponds to the square of the standardized autoregressive coefficient linking $occas_2$ to $occas_3$ (0.209^2). Thus, this autoregressive effect explains 4.4% of the total variance of $occas_3$, and not the total variance of S_DEP3, which we will now discuss.

```
lavaan (0.5-23.1097) converged normally after 196 iterations

                                                  Used       Total
  Number of observations                          3707        3777

  Number of missing patterns                       197

  Estimator                                         ML      Robust
  Minimum Function Test Statistic              262.942     189.919
  Degrees of freedom                               102         102
  P-value (Chi-square)                           0.000       0.000
  Scaling correction factor                                  1.384
    for the Yuan-Bentler correction

Model test baseline model:

  Minimum Function Test Statistic            19170.008   13515.928
  Degrees of freedom                               153         153
  P-value                                        0.000       0.000

User model versus baseline model:

  Comparative Fit Index (CFI)                    0.992       0.993
  Tucker-Lewis Index (TLI)                       0.987       0.990

  Robust Comparative Fit Index (CFI)                         0.994
  Robust Tucker-Lewis Index (TLI)                            0.990

Loglikelihood and Information Criteria:

  Loglikelihood user model (H0)             -78132.849  -78132.849
  Scaling correction factor                                  1.214
    for the MLR correction
  Loglikelihood unrestricted model (H1)     -78001.378  -78001.378
  Scaling correction factor                                  1.421
    for the MLR correction

  Number of free parameters                         87          87
  Akaike (AIC)                              156439.698  156439.698
  Bayesian (BIC)                            156980.662  156980.662
  Sample-size adjusted Bayesian (BIC)       156704.219  156704.219

Root Mean Square Error of Approximation:

  RMSEA                                          0.021       0.015
  90 Percent Confidence Interval          0.018  0.024       0.012  0.018
  P-value RMSEA <= 0.05                          1.000       1.000

  Robust RMSEA                                               0.018
  90 Percent Confidence Interval                             0.014  0.022

Standardized Root Mean Square Residual:

  SRMR                                           0.056       0.056
```

Table 4.6a. *Overall goodness-of fit indices from TSO applied to the CES-D scale*

Regressions:		Estimate	Std.Err	z-value	P(>|z|)	Std.lv	Std.all
occas2 ~							
occas1	(beta)	0.207	0.037	5.669	0.000	0.237	0.237
occas3 ~							
occas2	(beta)	0.207	0.037	5.669	0.000	0.209	0.209
occas4 ~							
occas3	(beta)	0.207	0.037	5.669	0.000	0.207	0.207
occas5 ~							
occas4	(beta)	0.207	0.037	5.669	0.000	0.207	0.207
occas6 ~							
occas5	(beta)	0.207	0.037	5.669	0.000	0.207	0.207

Table 4.6b. *"Regressions" rubric of TSO results applied to the CES-D scale (contd.)*

R-Square:	Estimate
S_DEP1	1.000
S_DEP2	1.000
S_DEP3	1.000
S_DEP4	1.000
S_DEP5	1.000
S_DEP6	1.000
occas2	0.056
occas3	0.044
occas4	0.043
occas5	0.043
occas6	0.043

Table 4.6c. *"R-Square" rubric of results for TSO applied to the CES-D scale (contd.)*

Variance decomposition of S_DEP factors

The following question is the rationale behind applying th TSO to the CES-D scale is: what captures the construct "depression" assessed by the CES-D scale administred at time *t*? Put simply, what assesses the construct operationalized (defined) through the CES-D scale: a depression-trait, a depression-state or both at once?

The TSO attempts to answer these questions, as it has the advantage of being able to partition the variance of the measure of the construct in only two parts: the part imputable to the stable trait and the part imputable to a specific occasion reflecting the state in which the respondent is at the assessment moment.

Variances:		Estimate	Std.Err	z-value	P(>|z|)	Std.lv	Std.all
S_DEP1		0.000				0.000	0.000
S_DEP2		0.000				0.000	0.000
S_DEP3		0.000				0.000	0.000
S_DEP4		0.000				0.000	0.000
S_DEP5		0.000				0.000	0.000
S_DEP6		0.000				0.000	0.000
occas1		4.895	0.306	16.020	0.000	1.000	1.000
.occas2	(zeta)	3.519	0.184	19.103	0.000	0.944	0.944
.occas3	(zeta)	3.519	0.184	19.103	0.000	0.956	0.956
.occas4	(zeta)	3.519	0.184	19.103	0.000	0.957	0.957
.occas5	(zeta)	3.519	0.184	19.103	0.000	0.957	0.957
.occas6	(zeta)	3.519	0.184	19.103	0.000	0.957	0.957
.CSD0DA		4.307	0.233	18.515	0.000	4.307	0.296
.CSD0PA		3.764	0.153	24.527	0.000	3.764	0.383
.CSD0SC		4.635	0.200	23.184	0.000	4.635	0.395
.CSD3DA		4.296	0.290	14.802	0.000	4.296	0.321
.CSD3PA		2.650	0.156	16.973	0.000	2.650	0.330
.CSD3SC		4.642	0.250	18.543	0.000	4.642	0.425
.CSD5DA		4.562	0.317	14.379	0.000	4.562	0.336
.CSD5PA		2.833	0.162	17.502	0.000	2.833	0.346
.CSD5SC		4.576	0.260	17.571	0.000	4.576	0.423
.CSD8DA		3.694	0.301	12.253	0.000	3.694	0.290
.CSD8PA		1.887	0.154	12.287	0.000	1.887	0.261
.CSD8SC		3.483	0.244	14.262	0.000	3.483	0.358
.CSD10DA		4.112	0.310	13.255	0.000	4.112	0.313
.CSD10PA		2.021	0.181	11.169	0.000	2.021	0.274
.CSD10SC		4.242	0.272	15.578	0.000	4.242	0.404
.CSD13DA		4.896	0.406	12.055	0.000	4.896	0.352
.CSD13PA		2.707	0.245	11.045	0.000	2.707	0.336
.CSD13SC		5.495	0.376	14.633	0.000	5.495	0.468
T_DEP		5.351	0.298	17.978	0.000	1.000	1.000

Table 4.6d. *"Variances" results rubrics for TSO applied to the CES-D scale (contd.)*

To do this, the "Variances" rubric in the results (Table 4.6d) offers us the beginnings of an answer. The construct's variance decomposition (S_DEP1) at the first instance of assessment is very clear. In fact,, the variance of the trait factor, Var(T_DEP) = 5.351, that of factor $occas_1$, Var(occas1) = 4.895, bringing a total variance of (5.351 + 4.895) 10.246. The part imputable to trait in the total variance of S_DEP1 is equal to (5.351/10.246 =) 0.522, which represent 52.2% of total variance. The remainder, which is 47.8%, is attributable to phenomena other than the trait, in this case to the specific context in which the respondent finds the respondent is during the assessment occasion ($occas_1$).

For the other assessment occasions, the variance decomposition is slightly different, as factors $occas_2$–$occas_6$ have no variance to estimate like $occas_1$, but each instead has a residual variable variance ($\zeta_2 - \zeta_6$ designated by "zeta" in the results table). This does not express the "occas" factor variance. The variances of ζ_2 to ζ_6 do not express the total variances of their respective factors ($occas_2$ to $occas_6$). The total variance of the "occas" contains also that of the autoregressive effect (designated by "beta" in the results). Thus, we must add the variance ζ_t (zeta) and the variance imputable to the autoregressive effect to obtain the total variance of each "occasion" ($occas_2$ to $occas_6$). For example, for $occas_2$, the autoregressive effect variance is obtained as follows: Var(occas1)*b² (where b = non-standardized coefficient $occas_1 \rightarrow occas_2$). The variance of $\zeta_2 = 3.519$ (see rubric "variances" in Table 4.6). For $occas_3$, the autoregressive effect variance is obtained as follows: Var(occas2)*b² (where b = non-standardized coefficient $occas_2 \rightarrow occas_3$). For $occas_6$, the autoregressive effect variance obtained is as follows: Var(occas5)*b² (where b = non-standardized coefficient $occas_5 \rightarrow occas_6$).

However, we can proceed differently and more simply to partition the variance of scores obtained on our scale at different times (S_DEP_t). Here is a demonstration based on factor loadings provided in the results rubric "Latent Variables" (Table 4.6).

We recall that at each assessment occasion *t*, scores (S_DEP_t) are affected by two separate sources of influence: the "stable trait" and the time-varying factor "occasion". And, as these two factors are orthogonal (not related to one another), their standardized loadings are equivalent to their respective correlations with factor S_DEP_t, and their squared standarized loadings indicate their respective proportions in the variance of S_DEP_t. For example, examining the "Latent Variables" results rubric shows that the standardized loading coefficient going from T_DEP to S_DEP1 is 0.723. Once squared, to give $0.723^2 = 0.522$, this value indicates that factor T_DEP was responsible for 52.2% of S_DEP1's total variance. The remaining 47.8% is attributable to factor $occas_1$. In fact, the standardized loading coefficient of this factor ($occas_1$) on the factor S_DEP1 is 0.691, which, once squared [$(0.691)^2 = 0.478$], represents 47.8%.

The standardized loading of S_DEP2 on T_DEP is .768, which, once squared ($0.768^2 = 0.589$), represents the proportion of T_DEP in the variance of S_DEP2, which is 58.9%. The standardized loading of S_DEP2 on $occas_2$ is 0.641, which once squared (0.411) indicates the proportion imputable to $occas_2$ in S_DEP2's variance, which is 41.10%. Thus, S_DEP2 owes 100% (58.9% + 41.10%) of its variance to these two factors.

The standardized loading of S_DEP3, S_DEP4, S_DEP5 and S_DEP6 on T_DEP is 0.770, which, once squared ($0.770^2 = 0.592$), represents the proportion of T_DEP in their respective variance, which is 59.20%. The standardized loading of S_DEP3, S_DEP4, S_DEP5 and S_DEP6 on their respective factors "occas" is 0.638, indicating once squared (0.408) the part imputable to "occas" in their respective variances, which is 40.80%. Thus, S_DEP3, S_DEP4, S_DEP5 and S_DEP6 owe 100% (59.20% + 40.80%) of their variance to these two factors.

It is surprising that there are different standardized loadings when these loadings had initially been fixed entirely at 1.00. This difference arises from the fact that the variables in question have different variances.

```
Parameter Estimates:

  Information                                  Observed
  Standard Errors                      Robust.huber.white

Latent Variables:
                    Estimate  Std.Err  z-value  P(>|z|)   Std.lv  Std.all
  S_DEP1 =~
    CSD0DA             1.000                               3.201    0.839
    CSD0PA     (a)     0.770    0.014   54.902    0.000    2.464    0.786
    CSD0SC     (b)     0.832    0.016   53.450    0.000    2.663    0.778
  S_DEP2 =~
    CSD3DA             1.000                               3.013    0.824
    CSD3PA     (a)     0.770    0.014   54.902    0.000    2.319    0.819
    CSD3SC     (b)     0.832    0.016   53.450    0.000    2.507    0.758
  S_DEP3 =~
    CSD5DA             1.000                               3.005    0.815
    CSD5PA     (a)     0.770    0.014   54.902    0.000    2.313    0.809
    CSD5SC     (b)     0.832    0.016   53.450    0.000    2.500    0.760
  S_DEP4 =~
    CSD8DA             1.000                               3.005    0.842
    CSD8PA     (a)     0.770    0.014   54.902    0.000    2.313    0.860
    CSD8SC     (b)     0.832    0.016   53.450    0.000    2.499    0.801
  S_DEP5 =~
    CSD10DA            1.000                               3.005    0.829
    CSD10PA    (a)     0.770    0.014   54.902    0.000    2.313    0.852
    CSD10SC    (b)     0.832    0.016   53.450    0.000    2.499    0.772
  S_DEP6 =~
    CSD13DA            1.000                               3.005    0.805
    CSD13PA    (a)     0.770    0.014   54.902    0.000    2.313    0.815
    CSD13SC    (b)     0.832    0.016   53.450    0.000    2.499    0.729
  T_DEP =~
    S_DEP1             1.000                               0.723    0.723
    S_DEP2             1.000                               0.768    0.768
    S_DEP3             1.000                               0.770    0.770
    S_DEP4             1.000                               0.770    0.770
    S_DEP5             1.000                               0.770    0.770
    S_DEP6             1.000                               0.770    0.770
  occas1 =~
    S_DEP1             1.000                               0.691    0.691
  occas2 =~
    S_DEP2             1.000                               0.641    0.641
  occas3 =~
    S_DEP3             1.000                               0.638    0.638
  occas4 =~
    S_DEP4             1.000                               0.638    0.638
  occas5 =~
    S_DEP5             1.000                               0.638    0.638
  occas6 =~
    S_DEP6             1.000                               0.638    0.638
```

Table 4.6e. *"Latent Variables" rubric (factor loadings) of the results for TSO applied on the CES-D scale (contd.)*

It is possible, though optional, to partition the variance of each factor from $occas_2$ to $occas_6$ into two parts: the proportions imputable to the autoregressive effect and the proportion attributable to the residual variable (ζ). The proportion attributable to the autoregressive effect is calculated by squaring its standardized coefficient and multiplying it by the proportion of variance of the "occas" factor it influences. For example, the autoregressive coefficient affecting $occas_2$ ($occas_1 \rightarrow occas_2$) is 0.237. If it is squared ($0.237^2 = 0.056$) and multiplied by the variance of $occas_2$ (0.411), to give 0.056*0.411 = 0.023, this shows that 0.023, that is 2.3% of the variance of $occas_2$ can be explained by the autoregressive effect. The remainder (0.411 – 0.023 = 0.388 or 41.10% – 2.30% = 38.8%), i.e. 38.80% of the variance of $occas_2$ is imputable to its residual error (ς2). Table 4.7 summarises the results for the other "occas". It will be noted that the part of variance imputable to the trait is weaker at t_1 (0.52) than at subsequent times (0.59) whereas in a TSO the size of this component is constant from one moment of assessment to another (Var(T_DEP) = 5.351). This difference is therefore explained by differences in the total variance at different assessment occasions. Thus, the proportion of variance attributable to the stable trait was weaker at t_1, as the total variance was higher at that moment. The first is calculated in relation to the second.

Parameter	S_DEP1	S_DEP2	S_DEP3	S_DEP4	S_DEP5	S_DEP6
Proportion of the trait in total variance	.522 (52.2%)	.589 (58.9%)	.592 (59.2%)	.592 (59.2%)	.592 (59.2%)	.592 (59.2%)
Proportion of the autoregressive effect	----	.023 (2.3%)	.018 (1.8%)	.018 (1.8%)	.018 (1.8%)	.018 (1.8%)
Proportion of the residual variable (ς)	----	.387 (38.7%)	.392 (39.2%)	.392 (39.2%)	.392 (39.2%)	.392 (39.2%)
Total proportion of "occas" factor	.478 (47.8%)	.411 (41.1%)	.408 (40.8%)	.408 (40.8%)	.408 (40.8%)	.408 (40.8%)
Standardized stability coefficient	----	.237	.209	.207	.207	.207

NOTE. – The proportion of factor "occasion" in S_DEP's variance is the sum of the proportion of the autoregressive effect and of that of the residual variable (ς). For example, 0.411 = 0.387 + 0.023. The sum of the proportion of the occasion factor and that of the trait represents 100% of S_DEP's variance. Similarly, the sum of the proportions of the trait factor, of the autoregressive effect and the residual variable represent 100% of S_DEP's variance. For example, the proportion of the autoregressive effect of $occas_2$ (= 0.023) is obtained as followed: $(0.237)^2$ *0.411. The proportion of the residual variable ς2 (= 0.387) is obtained as followed: 0.411 – 0.023 = 0.387.

Table 4.7. *Proportions (and %) of variance in scores on the depression scale (CES-D) explained by TSO's components*

4.2.3. *Concluding remarks*

Applied to depression scores obtained on six occasions using the CES-D scale, the TSO model makes it possible to "shed light" and remove doubts about the profound nature of the construct "depression" made operationalized by this scale. It would seem that scores on this scale capture the depression-trait (59%) rather than the depression-state (41%), even if this greater ability to capture the depression-trait remains fairly slight. Others might say that the scores on this scale capture depressiveness (trait) rather than depressive mood (state). It will also be noted that the autoregressive coefficients were fairly weak (between 0.21 and 0.24) although still statistically significant, indicating that situational and contextual influences on observed measure were not very stable over time.

We invite the reader to compare the results of STARTS and the TSO and to comment on the differences.

And to close this chapter, we make some recommendations that can be applied to all latent trait-state models.

First, testing a longitudinal measurement model is a requisite stage for applying trait-state models. This stage aims to estimate the measurement model using a longitudinal CFA to test a factorial model correlating all the repeated latent variables (see [LIU 17]). This longitudinal measurement model should also at least pass the weak and the strong measurement invariance tests, without which any real temporal change in the phenomenon measured by the construct would be confused with change over time in the measure itself. It is the measure that changes and not the phenomenon measured. A measure is called "invariable" when a score of the same value still represents the same quantity of the construct measured regardless of the moment at which it is assessed. In our example, the longitudinal measurement model contains six intercorrelated latent variables (one by assessment occasion) each defined by three indicators, including one fixed at 1.00 to guarantee their metrics, and two others, each constrained to equality over time. We recommend autocorrelating the same indicator's measurement errors over time to take account of the method's shared variance. Specification of such a model is as follows:

STEP 2. Specification of the longitudinal measurement model (CFA) with weak invariance.

```
model.SPE <-'
```

#1. Longitudinal CFAs (the latent variables are automatically intercorrelated).

```
S_DEP1 =~ 1*CSDODA + a*CSDOPA + b*CSDOSC
```

```
S_DEP2 =~ 1*CSD3DA + a*CSD3PA + b*CSD3SC
S_DEP3 =~ 1*CSD5DA + a*CSD5PA + b*CSD5SC
S_DEP4 =~ 1*CSD8DA + a*CSD8PA + b*CSD8SC
S_DEP5 =~ 1*CSD10DA + a*CSD10PA + b*CSD10SC
S_DEP6 =~ 1*CSD13DA + a*CSD13PA + b*CSD13SC
```

#2. Measurement error autocorrelations (45 correlations).

```
CSD0DA ~~ CSD3DA + CSD5DA + CSD8DA + CSD10DA +
CSD13DA
CSD3DA ~~ CSD5DA + CSD8DA + CSD10DA + CSD13DA
CSD5DA ~~ CSD8DA + CSD10DA + CSD13DA
CSD8DA ~~ CSD10DA + CSD13DA
CSD10DA ~~ CSD13DA
CSD0PA ~~ CSD3PA + CSD5PA + CSD8PA + CSD10PA +
CSD13PA
CSD3PA ~~ CSD5PA + CSD8PA + CSD10PA + CSD13PA
CSD5PA ~~ CSD8PA + CSD10PA + CSD13PA
CSD8PA ~~ CSD10PA + CSD13PA
CSD10PA ~~ CSD13PA
CSD0SC ~~ CSD3SC + CSD5SC + CSD8SC + CSD10SC +
CSD13SC
CSD3SC ~~ CSD5SC + CSD8SC + CSD10SC + CSD13SC
CSD5SC ~~ CSD8SC + CSD10SC + CSD13SC
CSD8SC ~~ CSD10SC + CSD13SC
CSD10SC ~~ CSD13SC'
```

STEP 3. Estimation of the longitudinal CFA.

```
model.EST <- cfa (model.SPE, data = BASE, missing =
"fiml", estimator = "MLR")
```

STEP 4. Retrieving the results of the TSO.

```
summary (model.EST, fit.measures = T, std = T,
rsq = T)
```

Then, proceed to estimate a standard TSO such as that shown in this illustration. Finally, if it is necessary and justified, test the different TSO with some constraints removed (i.e. the homogeneity of variances constraint or the homogeneity of autoregressive effects constraint) and compare their fit to the data with the standard TSO to retain the model that offers the best approximation of reality. For example, it might be thought that the "stable trait" factor could alone account for scores obtained at different assessment occasions at different moments of measurement and that the autoregressive effects are therefore useless. To demonstrate, it is enough to compare the fit of this model (without autoregressive paths, i.e. a second-order CFA model) with that of a standard TSO. Using $\Delta\chi^2$ along with Δdf, we can therefore verify the consequences of autoregressive effects on the model's overall fit (i.e. any consequence, improvement or deterioration).

4.3. Latent growth models

4.3.1. *General overview*

Development is intraindividual change. And as soon as change is mentioned, two axiomatic statements come to mind. First, everyone changes over time, but, and this is true of everyone, we do not change in the same way or at the same pace. For example, basic motor skills develop constantly between the ages of 0 and 9 years. However, children do not acquire these skills in the same way or at the same age. Therefore there are individual differences in the rhythm of this type of development and in the direction it may take. Secondly, change is a process that cannot be observed directly, it is therefore necessarily latent, but an approximation could be inferred by measuring its observable manifestations several times, that is longitudinally.

Longitudinal data is rare, precious and endlessly interesting to specialists in several sciences. The use of longitudinal data with the aid of models combining covariance structure analysis and mean structure modeling is a recent advance in SEM. It is called latent growth modeling (a generic name covering labels such as latent curve modeling, latent growth curve modeling, latent trajectory modeling) that makes it possible to consider changes over time affecting covariances, variances and means simultaneously. This technique makes it possible to describe an individual's initial level as well as their developmental trajectory (growth rate). It also makes it possible to evaluate interindividual variability in these trajectories. Finally, it offers the possibility of testing the effect of predictive variables on the initial level as well as on growth trajectories to determine some of their causes. A simple illustration is stated below.

Let us suppose that we have six repeated measures of depression in a sample of elderly people. The basic postulate that underpins latent growth modeling is that elderly people change in different ways over time. This change may be linear or non-linear (quadratic or exponential for example). Here, let us suppose that this change is a linear function of time:

$$y_{oi} = f_{oi} + f_{1i}(time) + \varepsilon_{oi} \qquad [4.3]$$

where:

– y_{oi} is an individuals' *i* depression score on one occasion o;

– f_{oi} is the intercept of each individual *i*, i.e. the individual's intial level assessed at first data collection (baseline or initial status);

– f_{1i} is the slope of each individual *i* depending on the occasion of assessment called "time", i.e. their growth rate;

– ε_{it} represents the error term (residual) related to the individual *i*'s score on occasion *o*, i.e. the discrepancy between the trajectory (estimated value) and the observed score (measure) at occasion *o* for this individual *i*.

In Ghisletta and McArdle [GHI 12] and in Preacher *et al.* [PRE 08b], the reader will find very didactic explanations of these models. In Curran, Obeidat and Losardo [CUR 10], they will find practical answers to frequently asked questions about them.

For each individual, we will now consider six repeated measures of depression (DEP), DEP1 at *t*0 (baseline), DEP2 at *t*1, DEP3 at *t*2, DEP4 at *t*3, DEP5 at *t*4 and DEP6 at *t*5; we also obtain the following system of equations:

Baseline: DEP1 = intercept + slope (0) + ε_1

Time1: DEP2 = intercept + slope (1) + ε_2

Time2: DEP3 = intercept + slope (2) + ε_3

Time3: DEP4 = intercept + slope (3) + ε_4

Time4: DEP5 = intercept + slope (4) + ε_5

Time5: DEP6 = intercept + slope (5) + ε_6

The originality of latent growth modeling lies in treating the intercept and slope as latent variables (factors). Development is thus considered to be a latent phenomenon. Thus, in the trajectory [4.3], y_{oi} is constitues for each individual into a sum of (1) an unobserved latent score (f_{oi}) representing its initial latent level (intercept), (2) an unobserved latent score (f_{1i}) representing its latent change over time (slope), and (3) the unobserved residual errors (also latent).

By respecting the basic conventions of modeling, the previous system of equations can be expressed using a structural model; Figure 4.4 provides an illustration of this in a graph form. A covariation between residual variables (e) is often admitted in this type of model. The shaded part in Figure 4.4 provides an illustration of it. This possibility is, in fact, one of the ways in which equation modeling is richer than classical models which, like ANCOVA, assume that error variances in repeated measures are equal and independent.

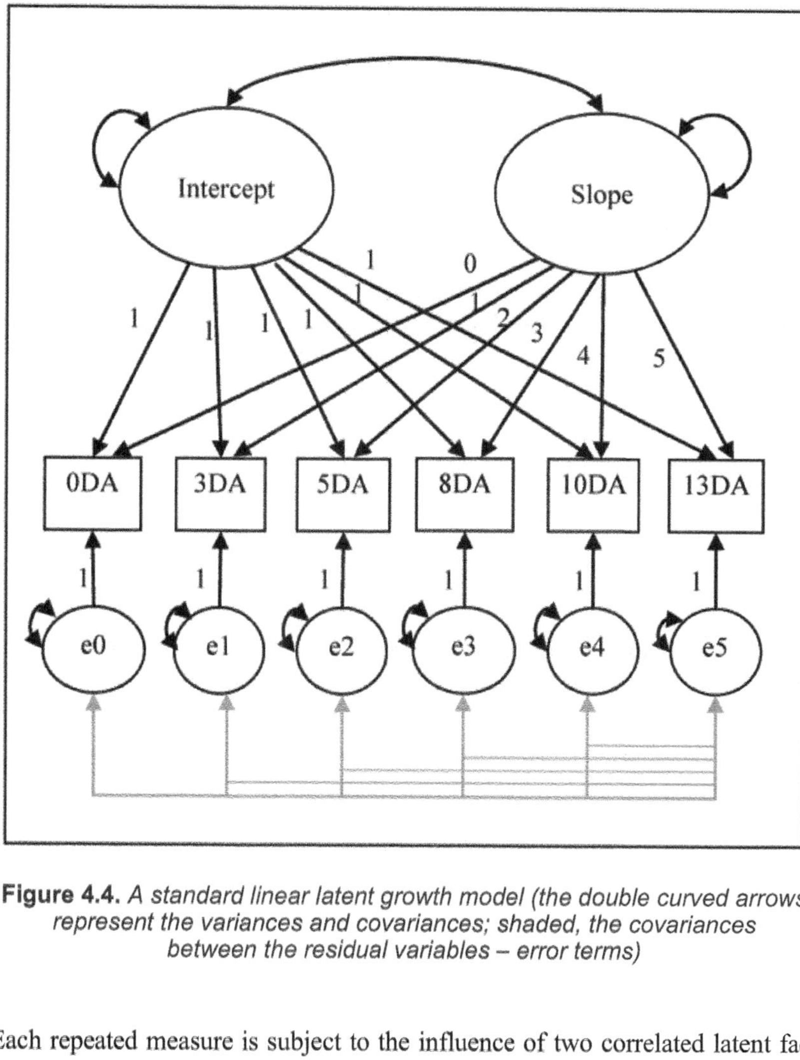

Figure 4.4. *A standard linear latent growth model (the double curved arrows represent the variances and covariances; shaded, the covariances between the residual variables – error terms)*

Each repeated measure is subject to the influence of two correlated latent factors. It will be noted that the paths expressing the direct weight of the intercept on these repeated measures are all fixed at 1.00, as the intercept factor represents the initial level (initial status) assessed as a baseline (sometimes called "time 0") and which remains invariable. For example, the depression score obtained as a baseline remains the same and will serve as a reference to estimate the change over time. However, the act of fixing paths expressing the weight of latent growth factor (slope) on the repeated measures means that we are more interested in these measures' variance and above all in the intercept and the slope variances. The intercept factor variance provides information on interindividual differences in the measure at baseline, while the slope factor variance provides information on interindividual differences relating to change over time. By fixing the weights of the (loadings) of the slope factor at 0 to 5, we hypothesize a linear change. The first weight (loading) is fixed at zero, as null change can affect the baseline score (the initial level will not change at all). Moreover, fixing a path (loading) at zero amounts quite simply to deleting it because it no longer has any weight.

The scaling of time is a crucial advantage in latent growth modeling. It expresses and specifies a priori the growth's pattern and nature. Here the researcher need to choose how to numerically code the passage of time. In fact, in the system of equations and in Figure 4.4 that illustrates it, time has been coded in such a way as to express the hypothesis of linear change. The factorial loading matrices (Λ) [4.4], [4.5] and [4.6] express this hypothesis:

$$\Lambda y = \begin{bmatrix} 1 & 0 \\ 1 & 1 \\ 1 & 2 \\ 1 & 3 \\ 1 & 4 \\ 1 & 5 \end{bmatrix} \qquad [4.4]$$

All the elements in the first column are fixed at 1.00 to express the fact that the intercept, that is the individual's intial level, remains constant across the assessments. The linear progression in the second column expresses the hypothesis of a linear change with equal time intervals. Change can be coded in different convertible time units chosen according to the research question: 0, 1, 2, 3, 4, 5, 6 years can be coded as 0, 12, 24, 36, 48, 60, 72 months.

The loading matrix [4.5], while still expressing the hypothesis that change is linear, codes time by respecting equal time intervals separating the assessment occasions (two units of time [weeks/months/years] separating the assessments from one another):

$$\Lambda y = \begin{bmatrix} 1 & 0 \\ 1 & 2 \\ 1 & 4 \\ 1 & 6 \\ 1 & 8 \\ 1 & 10 \end{bmatrix} \qquad [4.5]$$

However, obtaining assessments at equivalent intervals of time is not absolutely necessary for these models. The loading matrix [4.6] shows a slope (second column) with unequal time intervals: alternation between an interval with two units of time and an interval with three units of time between two experiments. In fact, when assessment occasions times are not equidistant, the time coding should take account of them and reflect the real intervals of time between these assessments. We emphasize however that the linear way in which time is coded does not affect the model's overall fit. Nonetheless, it influences and directs the way in which the model's results are interpreted:

$$\Lambda y = \begin{bmatrix} 1 & 0 \\ 1 & 2 \\ 1 & 5 \\ 1 & 7 \\ 1 & 10 \\ 1 & 13 \end{bmatrix} \qquad [4.6]$$

Similarly, in the absence of an initial hypothesis on the nature of the change, that is where there is an undetermined developmental trajectory, the nature of which we wish to explore, it is possible to free the time coding from the slope factor. Matrices [4.7] and [4.8] illustrate these suggestions. In [4.7] we read that the last three loadings of the last three assessment occasions are already free (*), and in [4.8] only the loading of the first assessment occasion and the loading of the last assessment occasion are fixed to explore a posteriori the shape of the growth trajectory of the phenomenon studied:

$$\Lambda y = \begin{bmatrix} 1 & 0 \\ 1 & 1 \\ 1 & 3 \\ 1 & * \\ 1 & * \\ 1 & * \end{bmatrix} \qquad [4.7]$$

$$\Lambda y = \begin{bmatrix} 1 & 0 \\ 1 & * \\ 1 & * \\ 1 & * \\ 1 & * \\ 1 & 5 \end{bmatrix} \qquad [4.8]$$

Latent growth models focus interest on intraindividual change. They focus on the nature of this change, that is on its rhythm and shape. Their use is however conditioned by at least three constraints. The first is the number of assessment occasions required to study the change, the minimum being three. In fact, where there are two assessment occasions, it is impossible to determine, in the universe of the possible and conceivable, the shape of the underlying developmental trajectory. It is clear that a multitude of different shapes (linear, polynomial) could theoretically cross both these assessment occasions. In addition, and it is more like a test-retest, a protocol with only two assessment occasions does not make it possible to distinguish real change from measurement error. The second addresses the psychometric properties of the measures used to assess the constructs under study. Temporal Measurement time-invariance is essential here, otherwise it is impossible to determine whether the observed change is imputable to variations in the measurement scale rather than to a real change in the phenomenon studied. This invariance should at least be strong (i.e. the temporal equality of the factor loadings and the measurement indicator intercepts). The third constraint addresses the nature of the phenomenon studied: is it a trait (or quasi-trait) or a state? Latent growth models are suitable for phenomena that are traits (or quasi-traits) whose change we wish to study. This is why a TSO could be a preliminary stage in a latent growth modeling.

We return for an instant to the graph in Figure 4.4 illustrating the standard latent growth model. It is a first-order univariate model. It is univariate as it only models change in a single construct (bivariate and multivariate versions exist, Figure 4.6 provides an illustration of them). It is a first-order model because it models observed measures of the construct and not latent constructs defined by multiple indicators. The graph in Figure 4.4 implicitly includes two important parameters, in this case the intercept mean and the slope mean. The graph in Figure 4.5 makes these explicit via the triangular constant 1.00 (see [MCA 14] for statistical details on this constant). The coefficients of path α_1 and α_2 are regressions on a constant, thus representing the intercept and slope factor means. The combination of two types of analysis, covariance structure analysis with lantent mean structure analysis, is a significant advantage of these models.

The key parameters of the model in Figure 4.5 are the elements called α_1 and V_1 for the latent variable intercept, and α_2 and V_2 for the latent variable slope. More precisely, α_1 represents the intercept factor mean and t V_1 the variance around this mean. When the latter proves to be statistically significant, this indicates that there are interindividual differences at the initial mean level of the phenomenon measured. In our example, this means that at the first data collection (at time 0), participants do not display the same level of depression. As for α_2, it represents the slope factor mean, that is the mean rate of change/growth in the participants over the course of the period studied, and V_2 the variance around this mean. When slope mean (α_2) proves to be statistically significant, this means that a substantial (non-null) intraindividual change has occurred. And when the variance around this mean proves to be statistically significant, we can infer the existence of interindividual differences in this intraindividual change (for example, participants do not change in the same way).

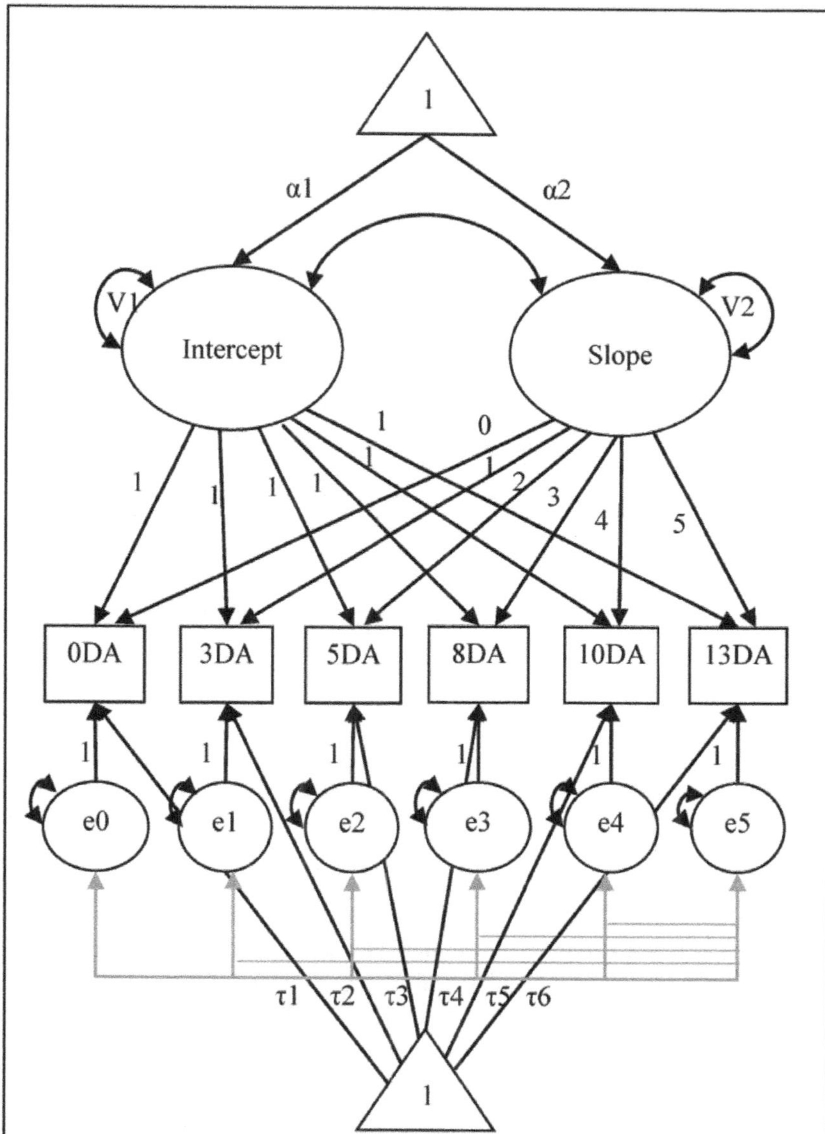

Figure 4.5. *Latent growth model where the means are explicit (the triangles represent constants, α1 = intercept factor mean, α2 = slope factor mean, τ1-τ6 = indicator intercepts; the double curved arrows represent the variances and covariances; shaded, the covariances between the residual variables)*

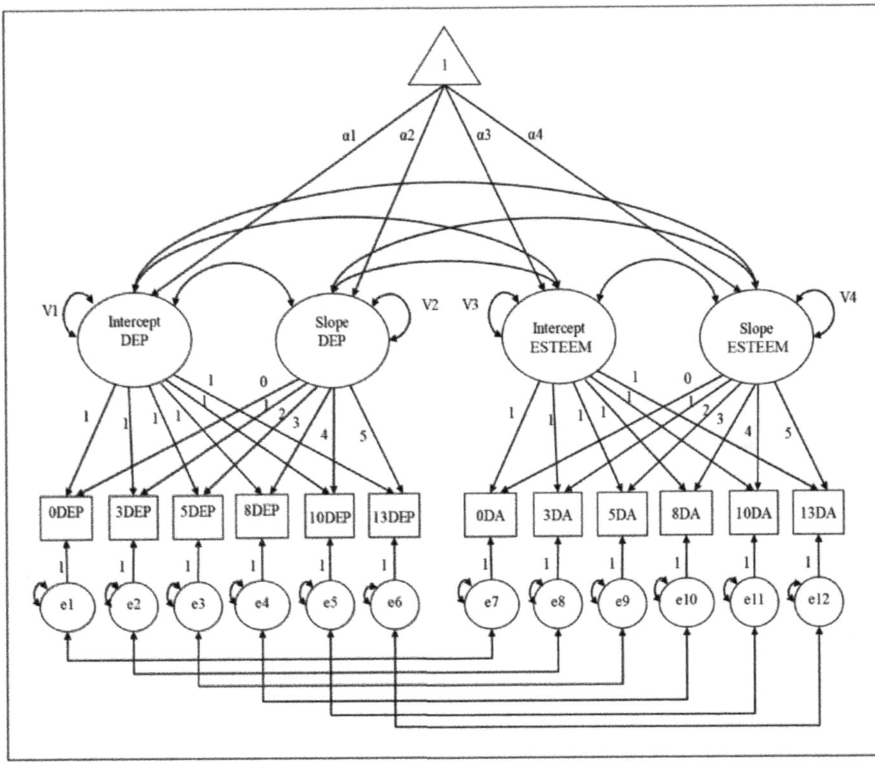

Figure 4.6. *Bivariate latent growth model (latent growth of two constructs: DEP = depression, ESTEEM = self-esteem) (the double curved arrows represent the variances and covariances; the covariances between the error terms are represented by straight double arrows; the triangle represents a constant)*

The correlation between the intercept and the slope indicates the relationship between the initial level and the change rate. It reveals the size of the upward or downward trend of the phenomenon studied. However, interpreting (this relationship is a delicate business, as it depends trend of growth (for example, an upward trend [positive slope mean] as opposed to a downward trend [negative slope mean]). When this relationship does not prove to be statistically significant, this suggests that the initial level does not predict the change in any way. A significant and positive relationship indicates that the initial high level predicts a substantial change in the phenomenon studied, whereas a negative relationship means that the initial high level predicts a lower level of intraindividual change. In other words, a high baseline level is linked to a low level of intraindividual change. For example, participants who displayed a high level of depression at the start of the study (at time 0) changed less markedly over the course of time.

If these main parameters reveal the existence of significant interindividual differences (above all concerning the slope, i.e. the change), it is useful if possible to explain why. Including explanatory (predictive) variables in a latent growth model could be an option. Figures 4.9, 4.11 and 4.12 each illustrate such an option, presenting a model called "conditional" as it includes some conditions (for example, the gender, the age at baseline) age likely to explain interindividual differences relating to intraindividual change. We will return to this later.

4.3.1.1. *Non-linear growth models*

Developmental trajectories cannot exist only as simple linear functions of time. Non-linear trajectories, that is ones that are not constant over time, are possible and can be envisaged. The most commonly used of these is the quadratic trajectory. Matrix [4.9] shows the loadings of a quadratic growth model. The first column represents the loadings on the intercept factor, the second the loadings on the slope factor modeling the linear change and the third column represents the loadings conveying the hypothesis of non-linear change. The loadings on the quadratic slope factor are the squares of the loadings on the linear slope factor. We have thus added a curve (growth acceleration or growth deceleration) to the linear component. Figure 4.7 illustrates this model in graph form:

$$\Lambda y = \begin{bmatrix} 1 & 0 & 0 \\ 1 & 1 & 1 \\ 1 & 2 & 4 \\ 1 & 3 & 9 \\ 1 & 4 & 16 \\ 1 & 5 & 25 \end{bmatrix} \qquad [4.9]$$

4.3.1.2. *Identification of a linear latent growth model*

First, at least three assessment occasions are needed to guarantee identification of the model. More assessment points are needed for complex (for example non-linear models). The free parameters to estimate in a standard latent growth model (figures 4.4 and 4.5) are: (1) two variances: the intercept variance and the slope variance, (2) two means: the intercept mean and the slope mean, (3) a covariance between the intercept and slope, and (4) variances in the error terms of repeated residual variable variances (the number of which depends on the number of assessment occasions).

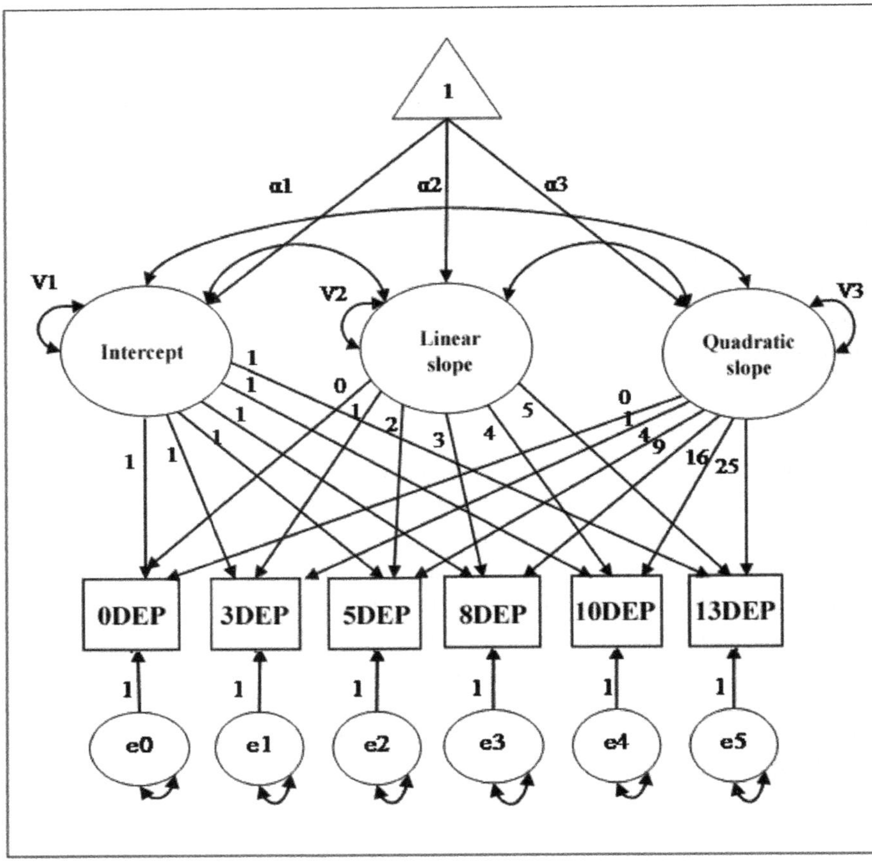

Figure 4.7. *Curvilinear latent growth model (the double curved arrows represent the variances and covariances; the triangle represents a constant)*

Other parameters, such as correlations between error/residual terms, can be estimated. And other constraints can be imposed, such as variance equality of error terms over time. We encourage the reader to draw this model accurately, detailing all the free parameters to be estimated, and finally to count the number of degrees of freedom using the following formula:

$$df = k - p$$

$$k = \frac{p * (p+3)}{2} \qquad [4.10]$$

where:

– k = the total number of information available (variances, covariances and means of the variables measured);

– p = number of free parameters to estimate

Note that the presence of means in the information counted makes these models true mean structure models.

4.3.1.3. *Specification of a linear latent growth model's parameters in lavaan syntax*

The specification of a standard univariate latent growth model (figures 4.4, 4.5) in lavaan syntax occurs along two simple lines (see the box below). The first involves the intercepts, considered to be a latent factor whose loadings are all set at 1.00. The second involves the slope, a latent change factor on which loaded each repeated measure, each obeying a linear time coding fixed by the researcher. Time zero defines the intercept, that is each participant's initial level (at time 0) in the phenomenon studied. The simplicity of this specification arises from the fact that lavaan offers a "growth" model-fitting function that automates the different details that latent growth models need for identification. For example, it will automatically fix the intercepts (estimated means) of the observed variables at zero.

4.3.2. *Illustration of an univariate linear growth model*

4.3.2.1. *Model specification*

For the purposes of our illustration, we return to the data used to present the univariate STARTS. This is a measure repeated six times, over a period of 13 years, with a dimension (i.e. Depressed Affect, DA) of the CES-D depression scale on a sample of elderly people. Applied to these data, the latent growth model aims mainly to examine intraindividual change over 13 years in this component of depression.

The following box presents the specification of this model in lavaan syntax (figures 4.4, 4.5). At STEP 2, there are two lines for model specification. The first involves the intercept factor. The second involves the slope factor whose time coding retains times intervals separating assessment occasions: zero represents the beginning, 3 means three years, 5 means five years, and 13 means 13 years after the baseline. At STEP 3, the stage reserved for commands on the model's estimation, note that a new "growth" model-fiiting function appears, which will explain the nature of the model to be estimated to the software.

STEP 2. Specification of the latent growth model (Figures 4.4, 4.5).

```
> model.SPE <- '
Intercept  =~  1*CSD0DA  +  1*CSD3DA  +  1*CSD5DA  +
1*CSD8DA + 1*CSD10DA + 1*CSD13DA
Slope  =~  0*CSD0DA  +  3*CSD3DA  +  5*CSD5DA  +
8*CSD8DA + 10*CSD10DA + 13*CSD13DA'
```

4.3.2.2. *Model evaluation*

STEP 3. Model estimation using "growth" model-fitting function.

```
model.EST <- growth (model.SPE, data = BASE, estimator
= "MLR", missing = "fiml")
```

STEP 4. Retrieving the results of the linear univariate latent growth model.

```
summary (model.EST, fit.measures = T, std = T)
```

It will be noted that the lavaan function for latent growth model is the model-fitting funxtion. A "robustness" indicator (MLR) has been retained to estimate the latent growth model. The missing data have been handled using the Full Information Maximum Likelihood (FIML) method.

4.3.2.2.1. Overall goodness-of-fit indices

We see from the results displayed on Table 4.8 that on the whole the data seem to tolerate the model well, as shown by the robust goodness-of-fit indices values: robust-CFI = 0.966, robust-TLI = 0.968, robust-RMSEA = 0.045, and this despite a statistically significant χ^2 for reasons we already know.

A brief word on the model's 16 degrees of freedom: this figure is obtained by subtracting the number of parameters to be estimated from the total number of available information. The latter one counts the measure's variances-covariances as well as their means. With 6 measured variables in the model, we count 21 variances-covariances (which is [6*(6 + 1)]/2) and 6 means (those of the 6 measured variables), which is a total of 27 pieces of information. The number of free parameters in the model is broken down as follows: 2 variances (the intercept variance and the slope variance) variance, 2 means (the intercept mean and the slope mean), 1 covariance between intercept and slope, and 6 variances (the error term variances), which is a total of 11 parameters to estimate. Thus, the number of degrees of freedom = (27 – 11) = 16.

```
lavaan (0.5-23.1097) converged normally after  54 iterations

                                                  Used       Total
  Number of observations                          3694        3777

  Number of missing patterns                        49

  Estimator                                         ML      Robust
  Minimum Function Test Statistic              143.939      88.803
  Degrees of freedom                                16          16
  P-value (Chi-square)                           0.000       0.000
  Scaling correction factor                                  1.621
    for the Yuan-Bentler correction

Model test baseline model:

  Minimum Function Test Statistic             3491.920    1835.419
  Degrees of freedom                                15          15
  P-value                                        0.000       0.000

User model versus baseline model:

  Comparative Fit Index (CFI)                    0.963       0.960
  Tucker-Lewis Index (TLI)                       0.966       0.963

  Robust Comparative Fit Index (CFI)                         0.966
  Robust Tucker-Lewis Index (TLI)                            0.968

Loglikelihood and Information Criteria:

  Loglikelihood user model (H0)             -29160.438  -29160.438
  Scaling correction factor                                  2.050
    for the MLR correction
  Loglikelihood unrestricted model (H1)     -29088.468  -29088.468
  Scaling correction factor                                  1.796
    for the MLR correction

  Number of free parameters                         11          11
  Akaike (AIC)                               58342.876   58342.876
  Bayesian (BIC)                             58411.235   58411.235
  Sample-size adjusted Bayesian (BIC)        58376.282   58376.282

Root Mean Square Error of Approximation:

  RMSEA                                          0.047       0.035
  90 Percent Confidence Interval          0.040  0.054       0.030  0.041
  P-value RMSEA <= 0.05                          0.782       1.000

  Robust RMSEA                                               0.045
  90 Percent Confidence Interval                             0.036  0.054

Standardized Root Mean Square Residual:

  SRMR                                           0.052       0.052
```

Table 4.8a. *Overall goodness-of fit indices for the latent growth model in Figure 4.4*

4.3.2.2.2. Local fit indices

First, it makes sense not to interpret the standardized estimations, except the covariance between the intercept and the slope.

Then, recall that the parameters of interest in a latent growth model are the intercept mean and above all the slope mean, their variances and finally the covariance between the intercept and the slope. These will be reviewed one by one, the continued part of Table 4.8.

```
Parameter Estimates:

  Information                                    Observed
  Standard Errors                      Robust.huber.white

Latent Variables:
                   Estimate  Std.Err  z-value  P(>|z|)
  Intercept =~
    CSD0DA            1.000
    CSD3DA            1.000
    CSD5DA            1.000
    CSD8DA            1.000
    CSD10DA           1.000
    CSD13DA           1.000
  Slope =~
    CSD0DA            0.000
    CSD3DA            3.000
    CSD5DA            5.000
    CSD8DA            8.000
    CSD10DA          10.000
    CSD13DA          13.000

Covariances:
                   Estimate  Std.Err  z-value  P(>|z|)
  Intercept ~~
    Slope            -0.155    0.042   -3.713    0.000

Intercepts:
                   Estimate  Std.Err  z-value  P(>|z|)
   .CSD0DA            0.000
   .CSD3DA            0.000
   .CSD5DA            0.000
   .CSD8DA            0.000
   .CSD10DA           0.000
   .CSD13DA           0.000
    Intercept         2.900    0.059   49.494    0.000
    Slope             0.010    0.008    1.284    0.199

Variances:
                   Estimate  Std.Err  z-value  P(>|z|)
   .CSD0DA            6.209    0.394   15.742    0.000
   .CSD3DA            6.017    0.395   15.245    0.000
   .CSD5DA            6.418    0.425   15.092    0.000
   .CSD8DA            4.877    0.385   12.671    0.000
   .CSD10DA           4.738    0.421   11.244    0.000
   .CSD13DA           7.303    0.755    9.674    0.000
    Intercept         8.057    0.442   18.247    0.000
    Slope             0.031    0.006    5.608    0.000
```

Table 4.8b. *Parameter estimates from the latent growth model in Figure 4.4 (contd.)*

The first parameters to consult are the intercept mean (initial status) and the slope mean (change rate), displayed in the "intercepts" rubric (here, indicating the latent variable means); the intercepts of the variables have been constrained to be null. Equal to 2.90, the intercept mean, that is depression mean level at the beginning of the study, is statistically significant. The same is not true of the slope mean (0.010) which, although positive, is not statistically significant ($p = 0.199$), thus indicating the absence of any linear change over a period of 13 years. Such a result opens the way for a hypothesis of non-linear change in depression, the justification for which should of course be supported by the literature.

However, the model has been modified by correlating residual variables adjacent to one another (section #2 in the box below), and by constraining their variances to be equal over time (section #1). The respecification of this model is shown in the box below.

STEP 2. Specification of the latent growth model (figures 4.4, 4.5) with constraints.

```
> model.SPE <- '
Intercept =~ 1*CSD0DA + 1*CSD3DA + 1*CSD5DA + 1*CSD8DA
+ 1*CSD10DA + 1*CSD13DA
Slope =~ 0*CSD0DA + 3*CSD3DA + 5*CSD5DA + 8*CSD8DA +
10*CSD10DA + 13*CSD13DA
```

#1. Optional parameters: constrain the residual variables' variances to equality over time.

```
CSD0DA ~~ v*CSD0DA
CSD3DA ~~ v*CSD3DA
CSD5DA ~~ v*CSD5DA
CSD8DA ~~ v*CSD8DA
CSD10DA ~~ v*CSD10DA
CSD13DA ~~ v*CSD13DA
```

#2. Optional parameters: correlate the adjacent residual variables.

```
CSD0DA ~~ CSD3DA
CSD3DA ~~ CSD5DA
CSD5DA ~~ CSD8DA
CSD8DA ~~ CSD10DA
CSD10DA ~~ CSD13DA'
```

STEP 3. Model estimation using the model-fiiting function "growth".

```
model.EST <- growth (model.SPE, data = BASE, estimator
= "MLR", missing = "fiml")
```

STEP 4. Retrieving the results of the univariate linear latent growth model.

```
summary (model.EST, fit.measures = T, std = T)
```

These modifications were not able to improve the model fit. The data collected among our participants does not seem to reflect a linear change in depression. Thus, the hypothesis of a non-linear change is becoming interesting. We therefore suggest testing this hypothesis.

4.3.3. *Illustration of an univariate non-linear (quadratic) latent growth model*

4.3.3.1. *Specification of the model*

The following box details specification of a latent growth model including a quadratic factor (slope factor) (Figure 4.7). Another line has been added. It involves the non-liner change factor (called "Slope Quadratic"), whose time coding represents the squares of the loadings associated with the linear slope factor.

The quadratic factor describes the upturn or downturn over time of the phenomenon studied beyond what is predicted by the linear factor [MUT 01]. Thus, the quadratic slope mean indicates the degree of quadratic curvature in the developmental trajectory [PRE 08b, SIN 03]. Here, the scaling of time respects the real chronology between assessment occasion.

STEP 2. Specification of the non-linear latent growth model (Figure 4.7, with a time coding respecting the interval between assessments).

```
> model.SPE <- '
Intercept =~ 1*CSD0DA + 1*CSD3DA + 1*CSD5DA + 1*CSD8DA
+ 1*CSD10DA + 1*CSD13DA
LinearSlope =~ 0*CSD0DA + 3*CSD3DA + 5*CSD5DA +
8*CSD8DA + 10*CSD10DA + 13*CSD13DA
SlopeQuadratic =~ 0*CSD0DA + 9*CSD3DA + 25*CSD5DA +
64*CSD8DA + 100*CSD10DA + 169*CSD13DA'
```

4.3.3.2. *Evaluation of the non-linear latent growth model*

STEP 3. Model estimation using the model-fiiting function "growth".

```
model.EST <- growth (model.SPE, data = BASE, estimator
= "MLR", missing = "fiml")
```

STEP 4. Retrieving the results of the non-linear latent growth model.

```
summary (model.EST, fit.measures = T, std = T)
```

4.3.3.2.1. Overall goodness-of-fit indices

According to the fit indices displayed in Table 4.9, and in comparison with those in Table 4.8, the non-linear latent growth model offers a better approximation of reality than the one hypothesizing a linear growth: robust-CFI = 0.987, robust-TLI = 0.983, robust-RMSEA = 0.032. In addition, the AIC and BIC values, often used in model comparison, were lower for the quadratic model, thus indicating that it fits the data better than the linear model.

4.3.3.2.2. The non-linear local fit indices

Let us begin by inspecting the "Intercepts" rubric, which offers the means of the models' three main latent factors (Table 4.11). It will be noted that these means are all statistically significant. Those of the linear slope factor and the quadratic slope factor are especially interesting, as they indicate, when they are statistically significant, that substantial changes (linear and non-linear) took place during the period covering the study.

It will be noted, and this point is not unimportant, that a linear slope mean (change) is negative (– 0.111) whereas the quadratic slope mean (change) is positive (0.011), indicating that there is a stage where depression decreases, preceding a stage where depression increases. We are witnessing an upward, concave growth curve (i.e. the growth curve is concave upward). The concavity inflection point, that is moment when the growth curve shifted to change direction, comes five years after the baseline (time 0).

This inflection point is calculated as follows [SIN 03]:

$$\frac{-M\_L}{2(M\_Q)} \qquad [4.11]$$

where:

- M_L = mean linear slope;
- M_Q = mean quadratic slope.

By applying this formula to our results, we obtain:

$$\frac{-(-0.111)}{2(0.011)} = 5.04$$

```
lavaan (0.5-23.1097) converged normally after 115 iterations

                                                  Used       Total
  Number of observations                          3694        3777

  Number of missing patterns                        49

  Estimator                                         ML      Robust
  Minimum Function Test Statistic               64.032      40.375
  Degrees of freedom                                12          12
  P-value (Chi-square)                           0.000       0.000
  Scaling correction factor                                  1.586
    for the Yuan-Bentler correction

Model test baseline model:

  Minimum Function Test Statistic             3491.920    1835.419
  Degrees of freedom                                15          15
  P-value                                        0.000       0.000

User model versus baseline model:

  Comparative Fit Index (CFI)                    0.985       0.984
  Tucker-Lewis Index (TLI)                       0.981       0.981

  Robust Comparative Fit Index (CFI)                         0.987
  Robust Tucker-Lewis Index (TLI)                            0.984

Loglikelihood and Information Criteria:

  Loglikelihood user model (H0)             -29120.484  -29120.484
  Scaling correction factor                                  1.963
    for the MLR correction
  Loglikelihood unrestricted model (H1)     -29088.468  -29088.468
  Scaling correction factor                                  1.796
    for the MLR correction

  Number of free parameters                         15          15
  Akaike (AIC)                               58270.969   58270.969
  Bayesian (BIC)                             58364.186   58364.186
  Sample-size adjusted Bayesian (BIC)        58316.523   58316.523

Root Mean Square Error of Approximation:

  RMSEA                                          0.034       0.025
  90 Percent Confidence Interval          0.026  0.043       0.019  0.032
  P-value RMSEA <= 0.05                          0.999       1.000

  Robust RMSEA                                               0.032
  90 Percent Confidence Interval                             0.021  0.043

Standardized Root Mean Square Residual:

  SRMR                                           0.037       0.037
```

Table 4.9. *Overall goodness-of-fit indices of the non-linear latent growth model*

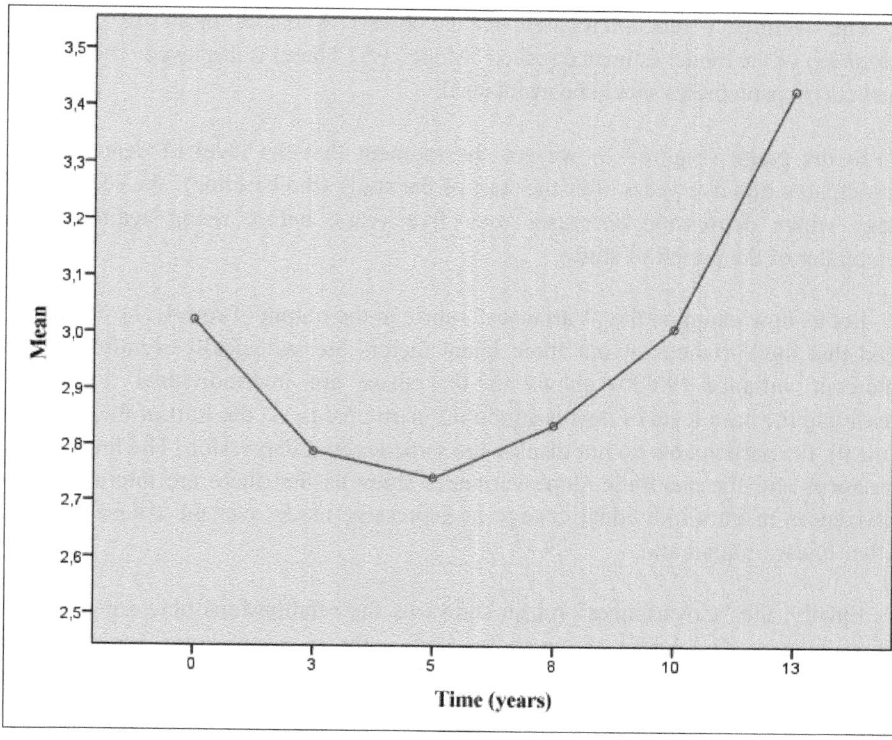

Figure 4.8. *Growth curve (estimated means) of a dimension of depression covering a period of 13 years*

An outline of this concave curve makes the results more eloquent. One can be made easily using *LibreOffice Calc*. Figure 4.8 is an example. The x-axis represents the time, coded in years, covering our variable's assessment occasions; our variable's estimated means are shown on the y-axis. These are obtained by requesting "mean.ov" of the estimated model (here model.EST) using the "lavInspect" function. Table 4.10 shows its specification and results:

```
> lavInspect (model.EST, "mean.ov")
 CSD0DA  CSD3DA  CSD5DA  CSD8DA CSD10DA CSD13DA
  3.022   2.789   2.742   2.837   3.009   3.430
```

Table 4.10. *Estimated means obtained with the "lavInspect" function*

The “lavInspect” function requests that the observed variables’ mean (ov = *observed variables*) of the model estimated (called “Model. EST” here) is displayed. The order of the items in parentheses should be maintained.

In the graph (Figure 4.8) we see the moment that the level of depression in participants tips five years after the start of the study (the baseline). We see a linear stage where depression decreases over five years, before rising again for the remainder of the period of study.

Let us now examine the “Variances” rubric in the output (Table 4.11). There, we read that the variances of our three latent factors are statistically significant. The intercept variance (9.620) shows us that there are interindividual differences involving the base level of depression in our participants. At the start of the study (at time 0), the participants do not display the same level of depression. The linear slope variances and the quadratic slope variances show us that there are interindividual differences in intra-individual change in depressive mode over the course of time, either linear or quadratic.

Finally, the “Covariances” rubric shows us the relationships between the three latent factors. To clarify as much to start with: in a quadratic latent growth model, the linear and quadratic functions are over partially overlapping and so are difficult to interpret separately. This is why it is recommended not to attach too much importance to their relationship.

We will conclude by emphasizing two points: (1) it can easily be said that a latent growth model with curvature provides a better description of our data than a linear model; (2) there are interindividual differences involving the developmental trajectories (linear and non-linear).

4.3.4. *Conditional latent growth model*

Where there are interindividual differences relating to intraindividual change, it is sometimes possible to find some of the reasons behind this. For example, could age at the beginning of the study (baseline), which can be highly variable (65 to 99 years in our sample), explain these differences? Including one or more predictive (measured and/or latent) variables (for example, age at baseline, gender, level of education, psychological variables) in a latent growth model makes it conditional latent growth model. Figures 4.9, 4.10 and 4.11 offer a presentation in graph form.

```
Latent Variables:
                    Estimate  Std.Err  z-value  P(>|z|)   Std.lv  Std.all
  Intercept =~
    CSD0DA             1.000                               3.102    0.829
    CSD3DA             1.000                               3.102    0.835
    CSD5DA             1.000                               3.102    0.830
    CSD8DA             1.000                               3.102    0.879
    CSD10DA            1.000                               3.102    0.868
    CSD13DA            1.000                               3.102    0.788
  LinearSlope =~
    CSD0DA             0.000                               0.000    0.000
    CSD3DA             3.000                               1.523    0.410
    CSD5DA             5.000                               2.538    0.679
    CSD8DA             8.000                               4.061    1.151
    CSD10DA           10.000                               5.076    1.421
    CSD13DA           13.000                               6.599    1.677
  QuadraticSlope =~
    CSD0DA             0.000                               0.000    0.000
    CSD3DA             9.000                               0.279    0.075
    CSD5DA            25.000                               0.774    0.207
    CSD8DA            64.000                               1.983    0.562
    CSD10DA          100.000                               3.098    0.867
    CSD13DA          169.000                               5.235    1.330

Covariances:
                    Estimate  Std.Err  z-value  P(>|z|)   Std.lv  Std.all
  Intercept ~~
    LinearSlope       -0.695    0.208   -3.334    0.001   -0.441   -0.441
    QuadraticSlope     0.033    0.014    2.390    0.017    0.346    0.346
  LinearSlope ~~
    QuadraticSlope    -0.015    0.005   -3.157    0.002   -0.940   -0.940

Intercepts:
                    Estimate  Std.Err  z-value  P(>|z|)   Std.lv  Std.all
   .CSD0DA             0.000                               0.000    0.000
   .CSD3DA             0.000                               0.000    0.000
   .CSD5DA             0.000                               0.000    0.000
   .CSD8DA             0.000                               0.000    0.000
   .CSD10DA            0.000                               0.000    0.000
   .CSD13DA            0.000                               0.000    0.000
    Intercept          3.022    0.061   49.180    0.000    0.974    0.974
    LinearSlope       -0.111    0.021   -5.322    0.000   -0.218   -0.218
    QuadraticSlope     0.011    0.002    6.405    0.000    0.352    0.352

Variances:
                    Estimate  Std.Err  z-value  P(>|z|)   Std.lv  Std.all
   .CSD0DA             4.364    0.664    6.569    0.000    4.364    0.312
   .CSD3DA             6.162    0.427   14.417    0.000    6.162    0.446
   .CSD5DA             6.272    0.464   13.511    0.000    6.272    0.449
   .CSD8DA             4.388    0.384   11.425    0.000    4.388    0.353
   .CSD10DA            4.579    0.409   11.204    0.000    4.579    0.359
   .CSD13DA            6.683    1.010    6.615    0.000    6.683    0.431
    Intercept          9.620    0.746   12.890    0.000    1.000    1.000
    LinearSlope        0.258    0.067    3.818    0.000    1.000    1.000
    QuadraticSlope     0.001    0.000    2.710    0.007    1.000    1.000
```

Table 4.11. *Parameter estimates from the non-linear latent growth model*

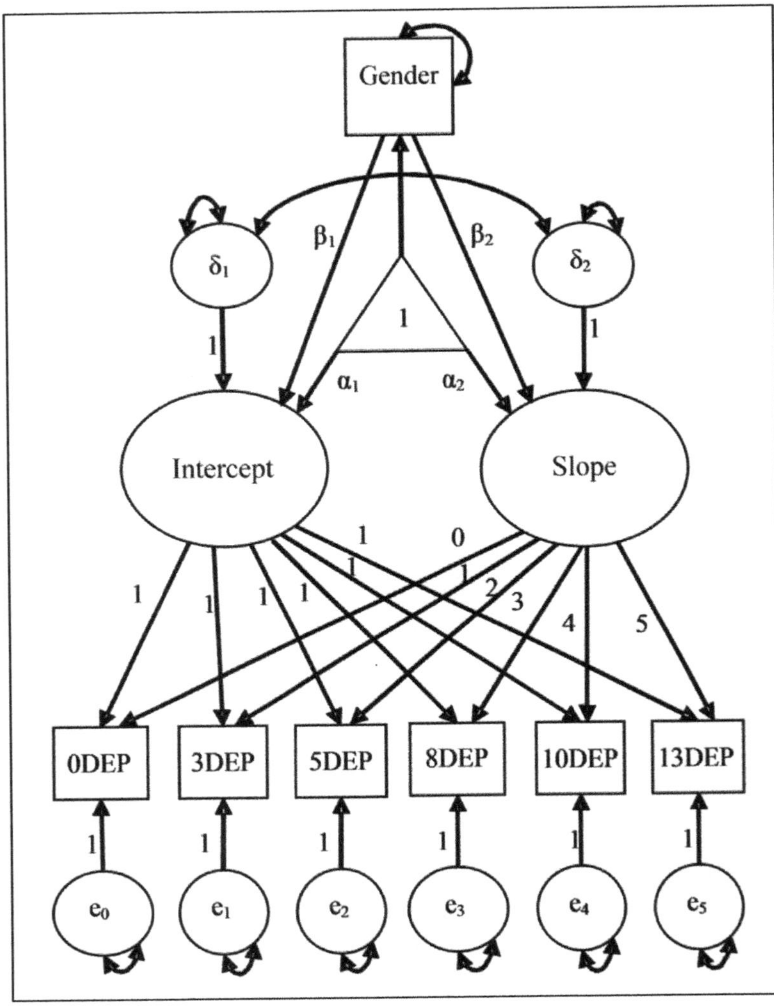

Figure 4.9. *Conditional growth model with time-invariant manifest covariate/ predictor (gender) (the double curved arrows represent the variances and covariances; the triangle represents a constant)*

In a latent growth modeling, two types of explanatory predictors can be included: (1) time-invariant predictors (for example, baseline age, gender, ethnic identity) whether they are manifest (Figure 4.9) or latent (Figure 4.11), (2) time-varying predictors (dynamic predictors) and for which there are repeated measures (for example age, self-esteem, subjective health) (Figure 4.10).

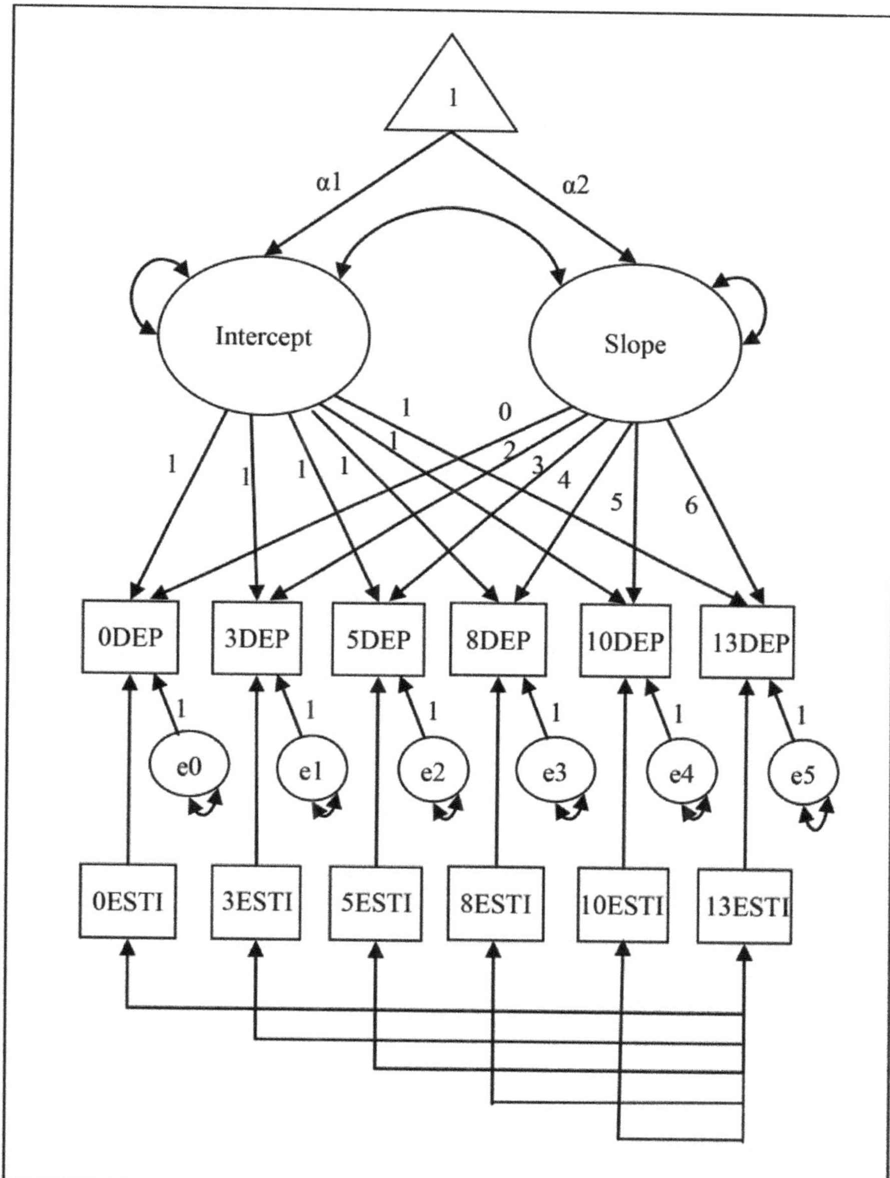

Figure 4.10. *Latent growth model with manifest time-varying covariates (ESTI) (the double curved arrows represent the variances and covariances; the covariances between covariates are represented by the double straight arrows; the triangle represents a constant)*

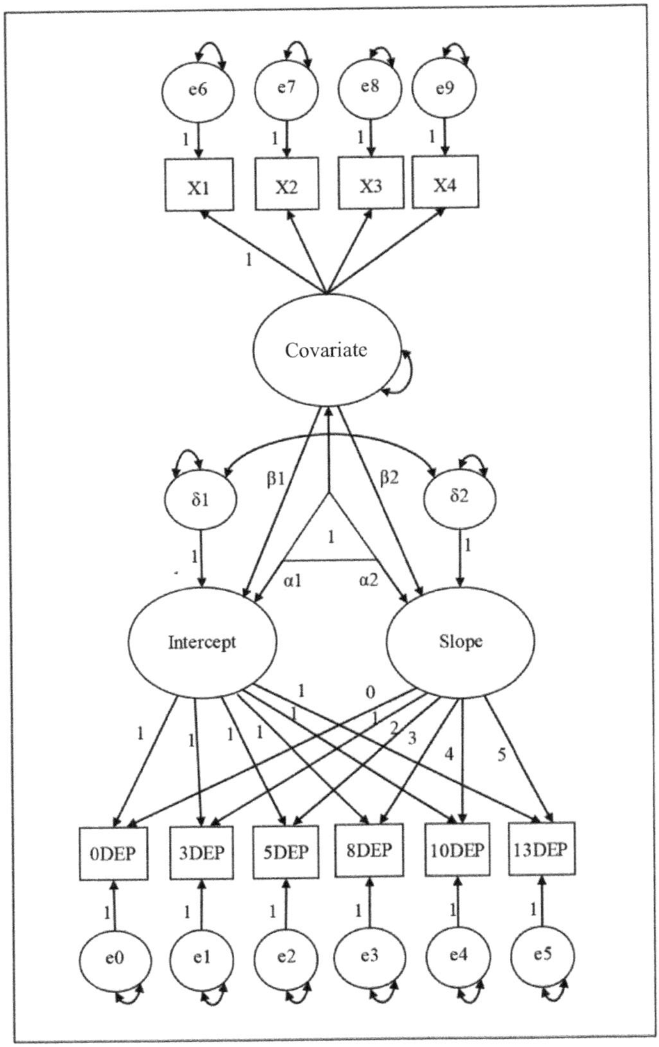

Figure 4.11. *Latent growth model with latent covariate measured by four indicators (the double curved arrows represent the variances and covariances)*

4.3.4.1. *Specification of a latent growth model's parameters with time-invariant covariates that are invariant with time*

Figure 4.12 includes the age of participants at the *baseline* as an An exogenous variable that predicts both the intercept and the two slopes. It will be noted that by becoming endogenous variables under the effect of the predictive variable, the latent factors (intercepts and slopes) lose their variances and their covariances, but each gain a residual/disturbance variable (ζ_1, ζ_2, ζ_3) whose variances and covariances will be estimated, reflecting the proportion of total variance that is not imputable to age at *baseline*. The parameters called "β" represent the effects of the predictor on the latent factors.

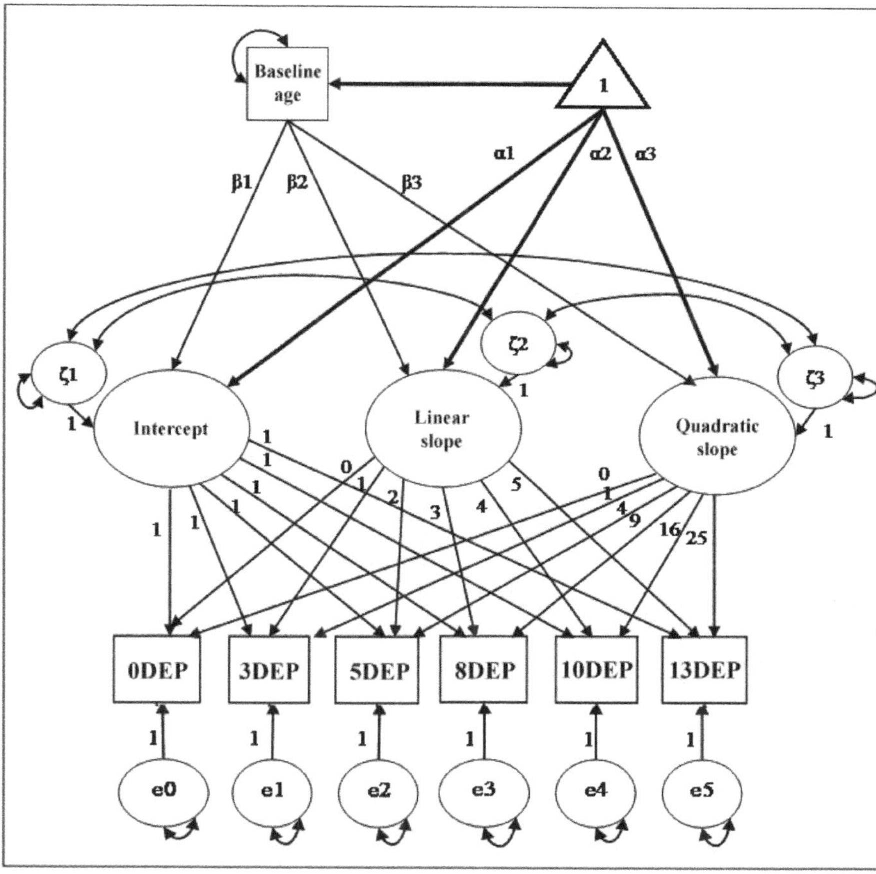

Figure 4.12. *Conditional non-linear growth model (the double curved arrows represent the variances and covariances; the triangle represents a constant)*

It makes sense to specify that the predictive (non-dichotomous) variables should be centered around their mean (which means subtracting the variable value from the mean of the same variable $(X - \overline{X})$) before including them in the model. Not centering these variables makes it difficult to interpret the intercept and slope means. STEP 1 in the box specifying the model on Figure 4.12 is reserved for commands, making it possible to center our predictive variable "age at baseline" (called "ageenter" in our BASE file) around the mean. The "attach ( )" function is needed to enable the variables in our data file to be read by R. The command "mean (ageenter)" makes it possible to verify that the variable has been well and truly centered.

STEP 1. Importing data.

```
> BASE <-read.csv2 (file = file.choose( ), sep = ";",
dec = ",")
```

Centering around the predictive variable mean "ageenter".

```
> attach (BASE)
> BASE$ageenter = ageenter  - mean (ageenter)
> mean (ageenter) # Option to verify that the variable
has been centered.
```

STEP 2. Specification of the conditional latent growth model (figure 4.12).

```
> model.SPE <- '
Intercept =~ 1*CSD0DA + 1*CSD3DA + 1*CSD5DA + 1*CSD8DA
+ 1*CSD10DA + 1*CSD13DA
LinearSlope =~ 0*CSD0DA + 3*CSD3DA + 5*CSD5DA +
8*CSD8DA + 10*CSD10DA + 13*CSD13DA
QuadraticSlope =~ 0*CSD0DA + 9*CSD3DA + 25*CSD5DA +
64*CSD8DA + 100*CSD10DA + 169*CSD13DA
Intercept ~ ageenter
LinearSlope ~ ageenter
QuadraticSlope ~ ageenter'
```

4.3.4.2. *Evaluation the conditional non-linear latent growth model's solution*

STEP 3. Model estimation using "growth" function.

```
model.EST <- growth (model.SPE, data = BASE, estimator
= "MLR", missing = "fiml")
```

STEP 4. Obtaining the solution of the conditional latent growth model.

```
summary (model.EST, fit.measures = T, std = T,
rsq = T)
```

Here, we will merely focus on the solution's local indices whose overall fit has proven to be excellent. Table 4.12 offers an extract of it. "Regressions" displays the effects of the ("ageenter") predictor on the model's three latent factors. Two effects have proven statistically significant and positive: the effect of baseline age on the intercept (β = 0.090, p = 0.000), indicating that the older the participant, the higher their level of depression at baseline, and the effect of baseline age on the linear slope (β = 0.123, p = 0.006), suggesting that the older the participant at baseline the higher their linear trajectory of depression. Baseline age had no significant effect on the curvature of the developmental trajectory (β = – 0.107, p = 0.100).

```
Regressions:
                     Estimate  Std.Err  z-value  P(>|z|)   Std.lv  Std.all
  Intercept ~
    ageenter            0.041    0.009    4.552    0.000    0.013    0.090
  LinearSlope ~
    ageenter            0.009    0.003    2.729    0.006    0.018    0.123
  QuadraticSlope ~
    ageenter           -0.000    0.000   -1.647    0.100   -0.015   -0.107

Covariances:
                     Estimate  Std.Err  z-value  P(>|z|)   Std.lv  Std.all
 .Intercept ~~
   .LinearSlope        -0.710    0.208   -3.414    0.001   -0.454   -0.454
   .QuadraticSlope      0.034    0.014    2.479    0.013    0.360    0.360
 .LinearSlope ~~
   .QuadraticSlope     -0.015    0.005   -3.159    0.002   -0.940   -0.940

Intercepts:
                     Estimate  Std.Err  z-value  P(>|z|)   Std.lv  Std.all
   .CSD0DA              0.000                               0.000    0.000
   .CSD3DA              0.000                               0.000    0.000
   .CSD5DA              0.000                               0.000    0.000
   .CSD8DA              0.000                               0.000    0.000
   .CSD10DA             0.000                               0.000    0.000
   .CSD13DA             0.000                               0.000    0.000
   .Intercept           3.026    0.061   49.377    0.000    0.974    0.974
   .LinearSlope        -0.091    0.022   -4.145    0.000   -0.178   -0.178
   .QuadraticSlope      0.010    0.002    5.479    0.000    0.334    0.334

Variances:
                     Estimate  Std.Err  z-value  P(>|z|)   Std.lv  Std.all
   .CSD0DA              4.340    0.663    6.547    0.000    4.340    0.310
   .CSD3DA              6.151    0.426   14.436    0.000    6.151    0.444
   .CSD5DA              6.289    0.464   13.547    0.000    6.289    0.448
   .CSD8DA              4.358    0.381   11.452    0.000    4.358    0.348
   .CSD10DA             4.616    0.412   11.204    0.000    4.616    0.357
   .CSD13DA             6.633    1.011    6.561    0.000    6.633    0.425
   .Intercept           9.563    0.747   12.806    0.000    0.992    0.992
   .LinearSlope         0.255    0.067    3.823    0.000    0.985    0.985
   .QuadraticSlope      0.001    0.000    2.712    0.007    0.989    0.989

R-Square:
                     Estimate
    CSD0DA              0.690
    CSD3DA              0.556
    CSD5DA              0.552
    CSD8DA              0.652
    CSD10DA             0.643
    CSD13DA             0.575
    Intercept           0.008
    LinearSlope         0.015
    QuadraticSlope      0.011
```

Table 4.12. *Parameter estimates for the conditional non-linear latent growth model (Figure 4.12)*

The results in the "R-Square" (R^2) rubric are worth commenting on. There, we read that participants' at the beginning of the study explains less than 1% of the intercept variance ($R^2 = 0.008$, which is 0.8%), explains 1.5% of linear change (slope) and 1.1% of variance in quadratic change. This means that other variables are needed to attempt to explain the interindividual differences relating to intraindividual change in our participants' depression.

4.3.5. *Second-order latent growth model*

It can never be emphasized enough that the major limitation of first order latent growth models lies in the fact that they analyze repeated measures that are assumed to be without measurement error. Just like the STARTS model, the univariate latent growth models that we have just introduced are not without this limitation. So, when data and sample size permit, it is preferable to opt for second-order models, that is those that include latent variable with multiple indicators. We will explore this option for latent growth models; Figure 4.13 offers an illustration in diagram form.

The model shown in Figure 4.13 is a hierarchical one. It is a second-order latent growth model. Each first-order latent factor "depression" (DEP1 to DEP6) has three indicators (i.e. three CES-D sub-dimensions, PA, DA and SC) assessed six times over a 13-year period. On each assessment occasion, each of the first order latent variables is defined by each indicator's intercept (estimated mean), the factor loading and the measurement error (in other words, each indicator in influenced by its own intercept, the factor weight, and measurement error). Generally, it will be noted that the same indicator's measurement errors are auto-correlated over time to take account of the shared method variance. The model hypothesizes a linear developmental trajectory. Thus, the intercept (initial status) and linear slope are the second-order factors (situated on two levels, below the indicators) and obeying the same constraints and time coding as those at work in the univariate linear models. Unlike the first-order univariate model, the intercept and slope do not act directly on the observed measures, but indirectly, via the latent variables, whose observed measures are the indicators. Just like the first order latent growth model, the intercept and slope means and variances are the key parameters of a second-order latent growth model. The advantage here, and it is a fairly significant one, is that estimations of such a model consider measurement errors relating to the indicators [CHA 98]. In addition, the second-order latent growth model captures change directly at the level of the construct itself rather than at the level of its indicators.

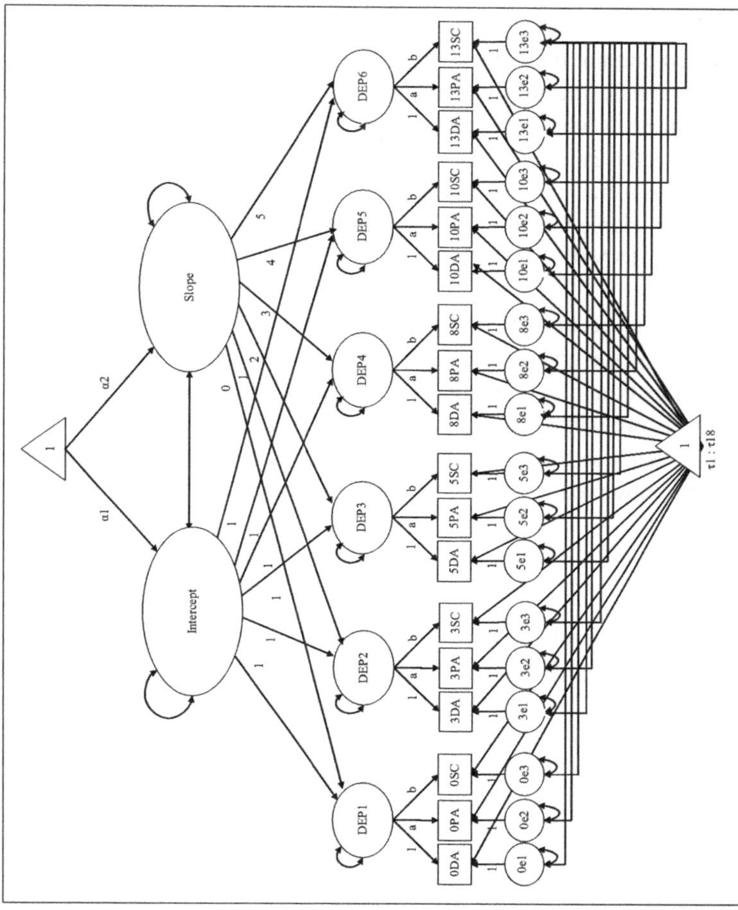

Figure 4.13. *Second-order latent growth model (the double curved arrows represent the variances; the triangles represent the constants; the covariances between the measurement errors are represented by the straight doubles arrows; τ1- τ18 = indicator intercepts)*

4.3.5.1. *Identification of a second-order latent growth model*

Because latent growth models are both models of covariance and mean structures, their identification is quite unique. The identification of firs-order factors is worth pausing over. Fixing a reference indicator loading at 1.00 to ensure a latent variable metric is an already known approach (covariance structure). For second order latent growth models, one should instead fix the reference indicator intercept at zero to ensure the metric of the latent variable's mean structure [HAN 01, STO 04]. We recall that a latent variable has no observed mean. Thus, on each assessment occasion, the metric of the first order factor mean will be based on the true, observed reference indicator mean and not on its intercept (i.e. the estimated mean), such that the second order latent variable means and variances (intercept and slope) will depend on the reference indicator's observed mean. Thus, the second order factors (intercept and slope) will directly capture the first order factor means, which are quite simply the reference indicator means. This is why it is not necessary to specify the estimation of the first order variable means since they automatically capture the reference indicator mean [CHA 98, HAN 01, LIT 13].

Moreover, the second-order latent growth model requires at least the strong hypothesis of time-invariance measurement. This hypothesis, having easily become verifiable in this type of model, guarantees that the construct will have been assessed on each occasion with the same measure. It also makes it possible to ensure that change occurs at the level of the construct (latent variable) rather than at the level of the observed variables (indicators) used to measure this construct [FER 08].

4.3.5.2. *Illustration of a second-order latent growth model*

We will use the model shown in Figure 4.13 as an illustration. It will be subject to the data already used to illustrate the TSO. At each assessment time, the latent variable "depression" is defined/assessed with three indicators, which are sub-dimensions of the CES-D scale.

4.3.5.2.1. Specification in lavaan syntax of a second-order latent growth model

The box below details the syntax that lavaan will need to estimate the model illustrated by Figure 4.13.

STEP 2. Specification of the second-order latent growth model (Figure 4.13).

```
model.SPE <- '
```

#1. First order latent "DEP" variables (weak invariance).

```
DEP1 =~ 1*CSD0DA + a*CSD0PA + b*CSD0SC
DEP2 =~ 1*CSD3DA + a*CSD3PA + b*CSD3SC
DEP3 =~ 1*CSD5DA + a*CSD5PA + b*CSD5SC
DEP4 =~ 1*CSD8DA + a*CSD8PA + b*CSD8SC
DEP5 =~ 1*CSD10DA + a*CSD10PA + b*CSD10SC
DEP6 =~ 1*CSD13DA + a*CSD13PA + b*CSD13SC
```

#2. Fixing the reference indicators' intercepts at zero .

```
CSD0DA ~ 0*1
CSD3DA ~ 0*1
CSD5DA ~ 0*1
CSD8DA ~ 0*1
CSD10DA ~ 0*1
CSD13DA ~ 0*1
```

#3. Do not estimate (i.e. fix at zero) the first-order latent variable means.

```
DEP1 ~ 0*1
DEP2 ~ 0*1
DEP3 ~ 0*1
DEP4 ~ 0*1
DEP5 ~ 0*1
DEP6 ~ 0*1
```

#4. Constrain the other indicators' intercepts to strong time equivalence (invariance).

```
CSD0PA ~ w*1
CSD0SC ~ z*1
CSD10PA ~ w*1
CSD10SC ~ z*1
CSD13PA ~ w*1
CSD13SC ~ z*1
CSD3PA ~ w*1
CSD3SC ~ z*1
CSD5PA ~ w*1
CSD5SC ~ z*1
CSD8PA ~ w*1
CSD8SC ~ z*1
```

#5. Measurement error autocorrelations (45 correlations, which is 15 for each indicator).

```
CSD0DA ~~ CSD3DA + CSD5DA + CSD8DA + CSD10DA + CSD13DA
CSD3DA ~~ CSD5DA + CSD8DA + CSD10DA + CSD13DA
CSD5DA ~~ CSD8DA + CSD10DA + CSD13DA
CSD8DA ~~ CSD10DA + CSD13DA
```

```
CSD0PA ~~ CSD3PA + CSD5PA + CSD8PA + CSD10PA + CSD13PA

CSD3PA ~~ CSD5PA + CSD8PA + CSD10PA + CSD13PA
CSD5PA ~~ CSD8PA + CSD10PA + CSD13PA
CSD8PA ~~ CSD10PA + CSD13PA
CSD10PA ~~ CSD13PA

CSD0SC ~~ CSD3SC + CSD5SC + CSD8SC + CSD10SC + CSD13SC
CSD3SC ~~ CSD5SC + CSD8SC + CSD10SC + CSD13SC
CSD5SC ~~ CSD8SC + CSD10SC + CSD13SC
CSD8SC ~~ CSD10SC + CSD13SC
CSD10SC ~~ CSD13SC
#6. Second-order factors of the linear model.
Intercept =~ 1*DEP1 + 1*DEP2 +1*DEP3 +1*DEP4 + 1*DEP5 +
1*DEP6
LinearSlope =~ 0*DEP1 + 3*DEP2 + 5*DEP3 + 8*DEP4 +
10*DEP5 + 13*DEP6'
```

The first five sections relate to specification of the parameters of the model's first-order factors. The strong measurement invariance is expressed by the invariance over time of both factor loadings (section #1) and indicators' intercepts loadings (section #4). The command in section #2 provides each first-order latent variable with a mean structure measure metric by fixing the intercept of their reference indicator, at zero. The mean of each first-order latent variable must also be fixed at zero (section #3), as this will be captured by the second order factors through the observed reference indicator mean. The last section (section #6) involves second-order factors, whose effects directly influence the first-order latent variables according to a linear time coding, comparable to that applied to the univariate model.

4.3.5.2.2. Evaluation of the second-order linear model

STEP 3. Model estimation using "growth" function.

```
model.EST <- growth (model.SPE, data = BASE, estimator
= "MLR", missing = "fiml")
```

STEP 4. Retrieving the results of the second order linear model.

```
summary (model.EST, fit.measures = T, std = T,
rsq = T)
```

Here too, a "robust" estimator (MLR) has been retained indicate to estimate the second-order latent growth model and the missing data have been handled using the Full Information Maximum Likelihood (FIML) method.

The model's overall fit has proven to be highly satisfactory. An inspection of local indices reveals that the slope mean was not statistically significant, suggesting the absence of linear change in depression. This result, conforming to that obtained using the previous first-order univariate model invites us to test a curvilinear latent growth model, in this case a quadratic one.

The specification of such a model is identical to that of the linear model with one exception, which is that the latent quadratic factor is added, provided with an appropriate time coding (section #6). The box below shows this addition.

#6. The second order factors of the curvilinear model (linear and quadratic coding).

```
Intercept =~ 1*DEP1 + 1*DEP2 +1*DEP3 +1*DEP4 + 1*DEP5 +
1*DEP6
LinearSlope =~ 0*DEP1 + 3*DEP2 + 5*DEP3 + 8*DEP4 +
10*DEP5 + 13*DEP6
QuadraticSlope =~ 0*DEP1 + 9*DEP2 + 25*DEP3 + 64*DEP4
+100*DEP5 + 169*DEP6
'
```

4.3.5.2.3. Evaluation of the solution of the curvilinear second-order model

STEP 3. Model estimation using "growth" model-fitting function.

```
model.EST <- growth (model.SPE, data = BASE, estimator
= "MLR", missing = "fiml")
```

STEP 4. Retrieving results of the second-order curvilinear model.

```
summary(model.EST, fit.measures = T, std = T, rsq = T)
```

Although χ^2 is significant (443.42, $df = 107$, $p = 0.000$), the other goodness-of-fit indices show that the model fits the data very well. Here, we will spare the reader, as far as possible, the irrelevant detail on a model's overall fit. Instead, we will draw their attention to the local indices, in this case the model's essential parameters.

The "intercepts" rubric, of which an extract is shown in Table 4.13, shows that the means of the three second-order latent factors are statistically significant at $p = 0.000$. This result indicates the existence of linear as well as curvilinear change.

Above all, it will be noted and has already been seen with the first-order univariate model, that the linear slope (change) mean is negative (–0.197) whereas that of the quadratic slope (change) is positive (0.018), suggesting that there is a stage where depression decreases, preceding a stage where it increases. We are seeing an upward concave growth curve.

```
Intercepts:

               Estimate  Std.Err  z-value  P(>|z|)

                 [...]

Intercept         3.175    0.063   50.646    0.000
LinearSlop       -0.197    0.019  -10.283    0.000
QuadrtcSlp        0.018    0.002   11.458    0.000
```

Table 4.13. *Extract from the "Intercepts" rubric, for the results of the curvilinear model displaying the means of three second-order factors*

It will be enough for the reader to apply [4.11] to obtain the inflection point, that is the moment where the growth curve tips, changing direction over the course of the period studied: this moment comes 5.47 years after the baseline [– (– 0.197)/2 (0.018) = 5.47 years]. Figure 4.14 makes it possible to visualize the developmental trajectory of depression as measured by the three dimensions of the CES-D scale over a 13-year period.

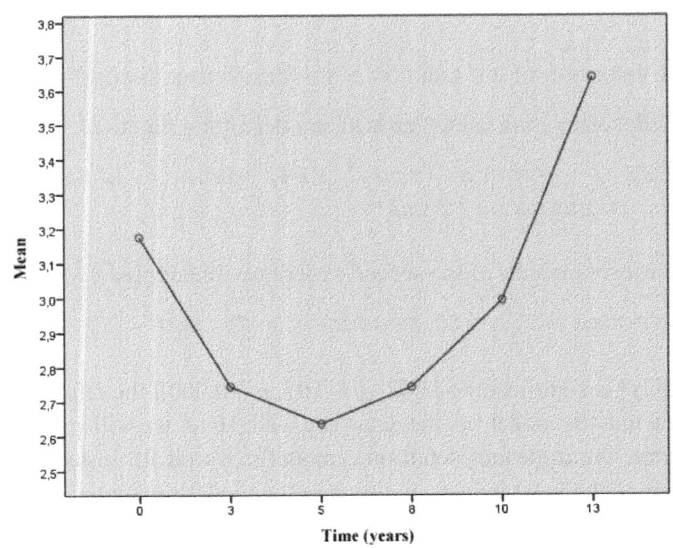

Figure 4.14. *Growth curve (estimated means) of depression over a period of 13 years*

The y-axis values on this figure are the first-order latent variable means (DEP1 to DEP6). Table 4.14 shows the command syntax to obtain them using the "lavInspect" function, as well as our model's results:

```
> lavInspect(model.EST, "mean.lv")
          DEP1           DEP2           DEP3           DEP4           DEP5           DEP6      Intercept
         3.175          2.744          2.636          2.743          2.993          3.638          3.175
   LinearSlope QuadraticSlope
        -0.197          0.018
```

Table 4.14. *Latent variable means for the second-order model obtained by using the "lavInspect" function*

The "lavInspect" function requests the restitution of the estimated model's (called "Model.EST" here) latent variable (lv = latent variables) means. It will be noted that the output offers the means of all our model's latent variables, that is the six first order latent variables (DEP1 – DEP6) as well as the three second-order latent variables (the intercept, the linear slope and the quadratic slope).

This figure's perfect resemblance to Figure 4.8 will not have escaped anyone. Nevertheless, the first illustrates a growth trajectory based on the scores of one of the three dimensions of the CES-D scale, while the second illustrates a trajectory based on the latent variable scores assessed by the three dimensions of the CES-D scale. This perfect resemblance is explained by the fact that the dimension used to illustrate the first-order univariate model (Figure 4.7) has been retained in the second-order model as a reference indicator whose intercept has been fixed at zero to offer the mean structure metric for each latent variable. On each assessment occasion, the latent variables have captured to the mean of this reference dimension. Hence the impossibility of reading and interpreting this mean as an absolute value. It is useful only to examine the evolution of the measured phenomenon over time, hence too the importance of choosing the reference indicator. We underline here, and this is fundamental and not remotely optional, that it is one of the limitations of the reference indicator strategy commonly used to identify a model. In Little [LIT 13] the reader will find a sustained critique of this strategy, as well as alternative strategies documented by the author.

The "Variances" rubric, an extract of which is shown in Table 4.15, shows that the variances around these means are statistically significant to $p = 0.00$, suggesting the existence of interindividual differences relating to intra-individual linear as well as curvilinear change in our participants' depression over a 13-year period.

```
Variances:

                    Estimate  Std.Err  z-value  P(>|z|)

                    [...]

Intercept             7.134    0.567   12.589    0.000
LinearSlope           0.173    0.050    3.485    0.000
QuadraticSlope        0.001    0.000    2.997    0.003
```

Table 4.15. *Extract from the "Variances" rubric for results of the curvilinear model displaying their variances around the means of three second-order factors*

It is possible, indeed desirable, to introduce time-invariant as well as time-varying covariates, measured as well as latent into a second-order growth model to attempt to explain these interindividual differences. These covariates will have the status of exogenous predictors affecting each of the three second-order latent variables (intercept, linear slope and quadratic slope).

We emphasize here that this way of predictor within latent growth models forms part of predictors is in line with a variable-oriented approach centered on the variables used to explain interindividual differences (heterogeneity) relating to intraindividual change. However, the heterogeneity of intraindividual change could be understood through an person-oriented approach centered on the individuals and not on the variables. For example, a multigroup analysis of latent growth makes it possible to explicitly examine change differences across predetermined groups (by gender, by age group, men versus women; young people versus adults). In absence of pre-selected subgroups, latent growth mixture modeling offers the possibilities of examining the heterogeneity of growth trajectories [MUT 99, NAG 99]. It is called "mixture" since it is a powerful and flexible technique combining latent growth analysis and latent class analysis. Far from being a recent development [GRE 51, LAZ 55], it is based on the general concept of categorical latent variables. It makes it possible to estimate an individual's membership of a discreet (unobserved) class from a set of variables. We define "class" as a sub-group of individuals situated at a at the same level in a latent variable. Integration of categorical latent variables within structural equation modeling where continuous latent variables predominate has gone some way to removing limitations more and has further extended the functions of equation modeling [MUT 02a]. Latent class growth modeling is a particular example of latent growth mixture modeling. So long as we can assume the existence of latent sources (classes) explaining the heterogeneity of developmental trajectories, it is preferable to resort to latent growth mixture modeling. This modeling hypothesizes that a population is thought to be formed of homogenous sub-groups (classes), each having its own specific growth trajectory. Identifying these latent homogenous classes, which are sources of heterogeneity within the study population, is the main objective of this type of modeling. In addition, within latent growth mixture modeling, data can be simultaneously analyzed through both variable and person-oriented approaches. For reasons of space, we will say that conceptually, latent class growth models aim to identify distinct sub-groups of individuals who have similar and homogenous growth profiles. These models thus highlight intraclass growth homogeneity and interclass growth heterogeneity. However, growth mixture modeling makes it possible to identify different latent classes (i.e. interclass growth heterogeneity) within which there may be individual differences (i.e. interclass growth heterogeneity).

We will not say anything further on this second extension provided by equation modeling. There are at least three reasons for this: first, this book has not tackled the logical basis for latent class models; second, the current version of lavaan does not allow these models to be estimated and finally, and above all because they merit it, there will be a monograph on them when their estimation is implemented, quite soon it seems, in lavaan.

4.4. Further reading

Readers who wish to know more about the main concepts addressed in this chapter can consult the following works:

BLALOCK H.J., *Causal Models in Experimental Designs*, Aldine Transaction, New Brunswick, 2009.

LITTLE T.D., *Longitudinal Structural Equation Modeling*, Guilford Press, New York, 2013.

MCARDLE J.J., NESSELROADE J.R., *Longitudinal Data Analysis Using Structural Equation Models*, American Psychological Association, Washington, 2014.

NEWSOM J.T., *Longitudinal Structural Equation Modeling: A Comprehensive Introduction*, Routledge/Taylor & Francis Group, New York, 2015.

PREACHER K.J., WICHMAN A.L., MACCALLUM R.C. *et al.*, *Latent Growth Curve Modeling*, Sage Publications, London, 2008.

References

[ABE 96] ABE S., BAGOZZI R.P., SADARANGANI P., "An investigation of construct validity and generalizability of the self-concept: self-consciousness in Japan and the United States", *Journal of International Consumer Marketing*, vol. 8, nos 3/4, pp. 97–123, 1996.

[ALD 77] ALDRICH J., "R.A. Fisher and the making of maximum likelihood 1912–1922", *Statistical Science*, vol. 12, pp. 162–176, 1977.

[ALL 36] ALLPORT G.W., ODBERT H.S., "Trait-names: a psycho-lexical study", *Psychological Monographs*, vol. 47, no. 1, pp. i–171, 1936.

[AND 88] ANDERSON J.C., GERBING D.W., "Structural equation modeling in practice: a review and recommended two-step approach", *Psychological Bulletin*, vol. 103, no. 3, pp. 411–423, 1988.

[ARB 95] ARBUCKLE J.L., *Amos for Windows. Analysis of Moment Structures (Version 3.5)*, Small Waters, Chicago, 1995.

[BAC 84] BACHER F., "Les méthodes statistiques en psychologie différentielle : perspective de développement", *Psychologie française*, vol. 29, pp. 9–15, 1984.

[BAC 87] BACHER F., "Les modèles structuraux en psychologie : Présentation d'un modèle : LISREL : I", *Le Travail humain*, vol. 50, no. 4, pp. 347–370, 1987.

[BAC 88] BACHER F., "Les modèles structuraux en psychologie : Présentation d'un modèle : LISREL : II", *Le Travail humain*, vol. 51, no. 4, pp. 273–288, 1988.

[BAC 99] BACHER F., "L'utilisation des modèles dans l'analyse des structures de covariance", *L'Année psychologique*, vol. 99, pp. 99–122, 1999.

[BAG 81] BAGOZZI R.P., FORNELL C., LARCKER D.F., "Canonical correlation analysis as a special case of a structural relations model", *Multivariate Behavioral Research*, vol. 16, no. 4, pp. 437–454, 1981.

[BAG 94] BAGOZZI R.P., HEATHERTON T.F., "A general approach to representing multifaceted personality constructs: application to state self-esteem", *Structural Equation Modeling*, vol. 1, pp. 35–67, 1994.

[BAG 98] BAGOZZI R.P., EDWARDS J.R., "A general approach for representing constructs in organizational research", *Organizational Research Methods*, vol. 1, pp. 45–87, 1998.

[BAR 86] BARON R.M., KENNY D.A., "The moderator-mediator variable distinction in social psychological research: conceptual, strategic, and statistical considerations", *Journal of Personality and Social Psychology*, vol. 51, no. 6, pp. 1173–1182, 1986.

[BAR 07] BARRETT P., "Structural equation modelling: adjudging model fit", *Personality and Individual Differences*, vol. 42, pp. 815–824, 2007.

[BEK 94] BEKKER P., "Counting rules for identification in linear structural models", *Computational Statistics & Data Analysis*, vol. 18, pp. 485–498, 1994.

[BEN 85] BENTLER P.M., Theory and Implementation of EQS, a Structural Equations Program, BMDP Statistical Software, Los Angeles, 1985.

[BEN 95] BENTLER P.M., EQS Structural Equations Program Manual, Multivariate Software, Encino, 1995.

[BEN 96] BENTLER P.M., DUDGEON P., "Covariance structure analysis: statistical practice, theory, and directions", *Annual Review of Psychology*, vol. 47, pp. 563–592, 1996.

[BEN 09] BENTLER P.M., "Alpha, dimension-free, and model-based internal consistency reliability", *Psychometrika*, vol. 74, pp. 137–143, 2009.

[BEN 10a] BENTLER P.M., "SEM with simplicity and accuracy", *Journal of Consumer Psychology*, vol. 20, no. 2, pp. 215–220, 2010.

[BEN 10b] BENTLER P.M., SATORRA A., "Testing model nesting and equivalence", *Psychological Methods*, vol. 15, no. 2, pp. 111–123, 2010.

[BER 89] BERNSTEIN I.H., TENG G., "Factoring items and factoring scales are different: spurious evidence for multidimensionality due to item categorization", *Psychological Bulletin*, vol. 105, no. 3, pp. 467–477, 1989.

[BER 07] BERGKVIST L., ROSSITER J.R., "The predictive validity of multiple-item versus single-item measures of the same constructs", *Journal of Marketing Research*, vol. 44, pp. 175–184, 2007.

[BLA 89] BLAIS M.R., VALLERAND R.J., PELLETIER L.G. *et al.*, "L'échelle de satisfaction de vie : validation canadienne-française du « Satisfaction with Life Scale »", *Revue canadienne des sciences du comportement*, vol. 21, no. 2, pp. 210–223, 1989.

[BLA 09] BLALOCK H.M., *Causal Models in Experimental Designs*, Aldine Transaction, New Brunswick, 2009.

[BOL 87] BOLLEN K.A., "Total, direct, and indirect effects in structural equation models", in CLOGG C.C. (ed.), *Sociological Methodology*, American Sociological Association, Washington, 1987.

[BOL 89] BOLLEN K.A., *Structural Equations with Latent Variables*, Wiley, New York, 1989.

[BOL 91] BOLLEN K.A., LENNOX R., "Conventional wisdom on measurement: a structural equation perspective", *Psychological Bulletin*, vol. 110, no. 2, pp. 305–314, 1991.

[BOL 93] BOLLEN K.A., LONG J.S., *Testing Structural Equation Models*, Sage Publications, New York, 1993.

[BOL 13] BOLLEN K.A., PEARL J., "Eight myths about causality and structural equation models", in MORGAN S. (ed.), *Handbook of Causal Analysis for Social Research*, Springer, London, 2013.

[BOO 82] BOOMSMA A., "Robustness of LISREL against small sample sizes in factor analysis models", in JORESKOG K.G., WOLD H. (eds), *Systems under Indirection Observation: Causality, Structure, Prediction (Part I)*, North Holland, Amsterdam, 1982.

[BOO 85] BOOMSMA A., "Non-convergence, improper solutions, and starting values in LISREL maximum likelihood estimation", *Psychometrika*, vol. 50, pp. 229–242, 1985.

[BOU 65a] BOUDON R., "A method of linear causal analysis: dependence analysis", *American Sociological Review*, vol. 30, pp. 365–374, 1965a.

[BOU 65b] BOUDON R., "Méthodes d'analyse causale", *Revue française de sociologie*, vol. VI, pp. 24–43, 1965b.

[BOU 67] BOUDON R., *L'Analyse mathématique des faits sociaux*, Plon, Paris, 1967.

[BOU 76] BOUDON R., "Comment on Hauser's review of education, opportunity, and social inequality", *American Journal of Sociology*, vol. 81, pp. 1175–1187, 1976.

[BRE 92] BRETON Y., "L'économie politique et les mathématiques en France, 1800–1940", *Histoire de la pensée économique*, pp. 25–52, 1992.

[BRO 84] BROWNE M.W., "Asymptotically distribution-free methods for the analysis of covariance structures", *British Journal of Mathematical and Statistical Psychology*, vol. 37, no. 1, pp. 62–83, 1984.

[BRO 89] BROWNE M.W., CUDECK R., "Single sample cross-validation indices for covariance structures", *Multivariate Behavioral Research*, vol. 24, no. 4, pp. 445–455, 1989.

[BRO 93] BROWNE M.W., CUDECK R., "Alternative ways of assessing model fit", in BOLLEN K.A., LONG J.S. (eds), *Testing Structural Equation Models*, Sage, Beverly Hills, 1993.

[BRO 15] BROWN T.A., *Confirmatory Factor Analysis for Applied Research*, 2nd edition, Guilford Press, New York, 2015.

[BRO 16] BROC G., *Stats faciles avec R*, De Boeck, Brussels, 2016.

[BUR 28] BURKS B.S., "The relative influence of nature and nurture upon mental development: a comparative study of foster parent-foster child resemblance and true parent-true child resemblance", *Yearbook of the National Society for the Study of Education*, pp. 219–316, 1928.

[BYR 89] BYRNE B.M., SHAVELSON R.J., MUTHÉN B., "Testing for the equivalence of factor covariance and mean structures: the issue of partial measurement invariance", *Psychological Bulletin*, vol. 105, no. 3, pp. 456–466, 1989.

[BYR 06] BYRNE B.M., *Structural Equation Modeling with EQS: Basic Concepts, Applications, and Programming*, 2nd edition, Lawrence Erlbaum Associates Publishers, Mahwah, 2006.

[CAT 47] CATTELL R.B., CATTELL A.S., RHYMER R.M., "P-technique demonstrated in determining psycho-physiological source traits in a normal individual", *Psychometrika*, vol. 12, pp. 267–288, 1947.

[CHA 98] CHAN D., "The conceptualization and analysis of change over time: an integrative approach incorporating longitudinal mean and covariance structures analysis (LMACS) and multiple indicator latent growth modeling (MLGM)", *Organizational Research Methods*, vol. 1, no. 4, pp. 421–483, 1998.

[CHE 06] CHEN F.F., WEST S.G., SOUSA K.H., "A comparison of bifactor and second order models of Quality of Life", *Multivariate Behavioral Research*, vol. 41, no. 2, pp. 189–225, 2006.

[CHE 08] CHEN F.F., CURRAN P.J., BOLLEN K.A. *et al.*, "An empirical evaluation of the use of fixed cutoff points in RMSEA test statistics in structural equation models", *Sociological Methods and Research*, vol. 36, pp. 462–494, 2008.

[CHO 95] CHOU C., BENTLER P.M., "Estimates and tests in structural equation modeling", in HOYLE R.H. (ed.), *Structural Equation Modeling: Concepts, Issues, and Applications*, Sage Publications, Thousand Oaks, 1995.

[COL 05] COLE D.A., MARTIN N.C., STEIGER J.H., "Empirical and conceptual problems with longitudinal trait-state models: introducing a trait-state-occasion model", *Psychological Methods*, vol. 10, no. 1, pp. 3–20, 2005.

[COM 92] COMREY A.L., LEE H.B., *A First Course in Factor Analysis*, 2nd edition, Lawrence Erlbaum Associates Publishers, Hillsdale, 1992.

[COR 93] CORTINA J.M., "What is coefficient alpha? An examination of theory and applications", *Journal of Applied Psychology*, vol. 78, pp. 98–104, 1993.

[CRO 51] CRONBACH L.J., "Coefficient alpha and the internal structure of tests", *Psychometrika*, vol. 16, pp. 297–334, 1951.

[CUD 83] CUDECK R., BROWNE M.W., "Cross-validation of covariance structures", *Multivariate Behavioral Research*, vol. 18, no. 2, pp. 147–167, 1983.

[CUD 89] CUDECK R., "Analysis of correlation matrices using covariance structure models", *Psychological Bulletin*, vol. 105, no. 2, pp. 317–327, 1989.

[CUR 10] CURRAN P.J., OBEIDAT K., LOSARDO D., "Twelve frequently asked questions about growth curve modeling", *Journal of Cognition and Development*, vol. 11, pp. 121–136, 2010.

[DEC 97] DECARLO L.T., "On the meaning and use of kurtosis", *Psychological Methods*, vol. 2, no. 3, pp. 292–307, 1997.

[DIA 08] DIAMANTOPOULOS A., RIEFLER P., ROTH K.P., "Advancing formative measurement models", *Journal of Business Research*, vol. 61, no. 12, pp. 1203–1218, 2008.

[DIC 94] DICKES P., TOURNOIS J., FLIELLER A. *et al.*, *La Psychométrie : théories et méthodes de la mesure en psychologie*, PUF, Paris, 1994.

[DIE 85] DIENER E., EMMONS R.A., LARSEN R.J. *et al.*, "The satisfaction with life scale", *Journal of Personality Assessment*, vol. 49, no. 1, pp. 71–75, 1985.

[DIL 87] DILLON W.R., KUMAR A., MULANI N., "Offending estimates in covariance structure analysis: comments on the causes of and solutions to Heywood cases", *Psychological Bulletin*, vol. 101, no. 1, pp. 126–135, 1987.

[DON 12] DONNELLAN M.B., KENNY D.A., TRZESNIEWSKI K.H. *et al.*, "Using trait-state models to evaluate the longitudinal consistency of global self-esteem from adolescence to adulthood", *Journal of Research in Personality*, vol. 46, no. 6, pp. 634–645, 2012.

[DUN 66] DUNCAN O.D., "Path analysis: sociological examples", *American Journal of Sociology*, vol. 72, no. 1, pp. 1–16, 1966.

[DUN 84] DUNCAN O.D., *Notes on Social Measurement, Historical and Critical*, Russell Sage Foundation, New York, 1984.

[EDW 00] EDWARDS J.R., BAGOZZI R.P., "On the nature and direction of relationships between constructs and measures", *Psychological Methods*, vol. 5, no. 2, pp. 155–174, 2000.

[END 11] ENDERS C.K., "Analyzing longitudinal data with missing values", *Rehabilitation Psychology*, vol. 56, no. 4, pp. 267–288, 2011.

[EYS 83] EYSENCK H.J., "Cicero and the state-trait theory of anxiety: another case of delayed recognition", *American Psychologist*, vol. 38, no. 1, pp. 114–115, 1983.

[FAB 99] FABRIGAR L.R., WEGENER D.T., MACCALLUM R.C. *et al.*, "Evaluating the use of exploratory factor analysis in psychological research", *Psychological Methods*, vol. 4, pp. 272–299, 1999.

[FAB 12] FABRIGAR L.R., WEGENER D.T., *Exploratory factor analysis*, Oxford University Press, New York, 2012.

[FAV 72] FAVERGE J.-M., *Méthodes statistiques en psychologie appliquée*, vol. 2, PUF, Paris, 1972.

[FER 08] FERRER E., BALLUERKA N., WIDAMAN K.F., "Factorial invariance and the specification of second-order latent growth models", *Methodology: European Journal of Research Methods for the Behavioral and Social Sciences*, vol. 4, no. 1, pp. 22–36, 2008.

[FIN 13] FINNEY S.J., DISTEFANO C., "Non-normal and categorical data in structural equation modeling", in HANCOCK G.R., MUELLER R.O. (eds), *Structural Equation Modeling: A Second Course*, 2nd edition, IAP Information Age Publishing, Charlotte, 2013.

[FIS 25] FISHER R.A., *Statistical Methods for Research Workers*, Oliver & Boyd, Edinburgh, 1925.

[FIS 37] FISHER R.A., "Professor Karl Pearson and the method of moments", *Annals of Eugenics*, vol. 7, no. 4, pp. 303–318, 1937.

[FLO 04] FLORA D.B., CURRAN P.J., "An empirical evaluation of alternative methods of estimation for confirmatory factor analysis with ordinal data", *Psychological Methods*, vol. 9, no. 4, pp. 466–491, 2004.

[FOR 81] FORNELL C., LARCKER D.F., "Evaluating structural equation models with unobservable variables and measurement error", *Journal of Marketing Research*, vol. 18, pp. 39–50, 1981.

[FOR 92a] FORNELL C., YI Y., "Assumptions of the two-step approach to latent variable modeling", *Sociological Methods & Research*, vol. 20, no. 1, pp. 291–320, 1992a.

[FOR 92b] FORNELL C., YI Y., "Assumptions of the two-step approach: reply to Anderson and Gerbing", *Sociological Methods & Research*, vol. 20, no. 1, pp. 334–339, 1992b.

[FOX 06] FOX J., "Structural equation modeling with the sem package in R", *Structural Equation Modeling*, vol. 13, pp. 465–486, 2006.

[FRE 87] FREEDMAN D.A., "As others see us: a case study in path analysis", *Journal of Educational Statistics*, vol. 12, no. 2, pp. 101–128, 1987.

[FRI 33] FRISCH R., WAUGH F.V., "Partial time regressions as compared with individual trends", *Econometrica*, vol. 1, pp. 387–401, 1933.

[FRI 86] FRIDHANDLER B.M., "Conceptual note on state, trait, and the state-trait distinction", *Journal of Personality and Social Psychology*, vol. 50, no. 1, pp. 169–174, 1986.

[GAN 13] GANA K., SAADA Y., BAILLY N. *et al.*, "Longitudinal factorial invariance of the Rosenberg self-esteem scale: determining the nature of method effects due to item wording", *Journal of Research in Personality*, vol. 47, no. 4, pp. 406–416, 2013.

[GEN 76] GENDRE F., *L'Analyse statistique multivariée. Introduction à son utilisation pratique*, Librairie Droz, Geneva, 1976.

[GHI 12] GHISLETTA P., MCARDLE J.J., "Teacher's corner: latent curve models and latent change score models estimated in R", *Structural Equation Modeling*, vol. 19, no. 4, pp. 651–682, 2012.

[GOL 72] GOLDBERGER A.S., "Structural equation methods in the social sciences", *Econometrica*, vol 40, pp. 979–1001, 1972.

[GOL 73] GOLDBERGER A.S., DUNCAN O.D., *Structural Equation Models in the Social Sciences*, Academic Press, New York, 1973.

[GON 99] GONZALEZ R., GRIFFIN D., "The correlational analysis of dyad-level data in the distinguishable case", *Personal Relationships*, vol. 6, pp. 449–469, 1999.

[GRA 80] GRANGER C.W.J., "Testing for causality. A personal viewpoint", *Journal of Economic Dynamics and Control*, vol. 2, pp. 329–352, 1980.

[GRA 05] GRACE J.B., BOLLEN K.A., "Interpreting the results from multiple regression and structural equation models", *Bulletin of the ESA*, vol. 86, pp. 283–295, 2005.

[GRA 08] GRACE J.B., BOLLEN K.A., "Representing general theoretical concepts in structural equation models: the role of composite variables", *Environmental and Ecological Statistics*, vol. 15, pp. 191–213, 2008.

[GRA 09] GRAHAM J.W., "Missing data analysis: making it work in the real world", *Annual Review of Psychology*, vol. 60, pp. 549–576, 2009.

[GRE 51] GREEN B.J., "A general solution for the latent class model of latent structure analysis", *Psychometrika*, vol. 16, pp. 151–166, 1951.

[GRE 82] GREENBERG D.F., KESSLER R.C., "Equilibrium and identification in linear panel models", *Sociological Methods & Research*, vol. 10, pp. 435–451, 1982.

[GRE 04] GREWAL R., COTE J.A., BAUMGARTNER H., "Multicollinearity and measurement error in structural equation models: implications for theory testing", *Marketing Science*, vol. 23, no. 4, pp. 519–529, 2004.

[GRI 95] GRIMM L.G., YARNOLD R., *Reading and understanding multivariate statistics*, APA, Washington, 1995.

[HAI 17] HAIR J.J., BABIN B.J., KREY N., "Covariance-based structural equation modeling in the Journal of Advertising: review and recommendations", *Journal of Advertising*, vol. 46, no. 1, pp. 163–177, 2017.

[HAN 01] HANCOCK G.R., KUO W.-L., LAWRENCE F.R., "An illustration of second-order latent growth models", *Structural Equation Modeling*, vol. 8, no. 3, pp. 470–489, 2001.

[HAU 76] HAUSER R.M., "On Boudon's model of social mobility", *American Journal of Sociology*, vol. 81, pp. 911–928, 1976.

[HAY 96] HAYDUK L.A., *LISREL: Issues, Debates, and Strategies*, Johns Hopkins University Press, Baltimore, 1996.

[HER 87] HERTZOG C., NESSELROADE J.R., "Beyond autoregressive models: some implications of the trait-state distinction for the structural modeling of developmental change", *Child Development*, vol. 58, no. 1, pp. 93–109, 1987.

[HER 13] HERSHBERGER S.L., MARCOULIDES G.A., "The problem of equivalent structural models", in HANCOCK G.R., MUELLER R.O. (eds), *Structural Equation Modeling: A Second Course*, 2nd edition, IAP Information Age Publishing, Charlotte, 2013.

[HOW 98] HOWELL D.C., *Méthodes statistiques en sciences humaines*, DeBoeck, Brussels, 1998.

[HOY 95] HOYLE R.H., PANTER A.T., "Writing about structural equation models", in HOYLE R.H. (ed.), *Structural Equation Modeling: Concepts, Issues, and Applications*, Sage Publications, Thousand Oaks, 1995.

[HU 95] HU L., BENTLER P.M., "Evaluating model fit", in HOYLE R.H. (ed.), *Structural Equation Modeling: Concepts, Issues, and Applications*, Sage Publications, Thousand Oaks, 1995.

[HU 98] HU L., BENTLER P.M., "Fit indices in covariance structure modeling: sensitivity to underparameterized model misspecification", *Psychological Methods*, vol. 3, no. 4, pp. 424–453, 1998.

[HU 99] HU L., BENTLER P.M., "Cutoff criteria for fit indexes in covariance structure analysis: conventional criteria versus new alternatives", *Structural Equation Modeling*, vol. 6, no. 1, pp. 1–55, 1999.

[JAC 96] JACCARD J., WAN C.K., *LISREL Approaches to Interaction Effects in Multiple Regression*, Sage Publications, Thousand Oaks, 1996.

[JAC 03] JACKSON D.L., "Revisiting sample size and number of parameter estimates: some support for the N: q hypothesis", *Structural Equation Modeling*, vol. 10, pp. 128–141, 2003.

[JAM 82] JAMES L.R., MULAIK S.A., BRETT J.M., *Causal Analysis: Assumptions, Models and Data*, Sage, Beverly Hills, 1982.

[JÖR 67] JÖRESKOG K.G., GRUVAEUS G., "RMLFA – A computer program for restricted maximum likelihood factor analysis", *Research Memorandum*, pp. 67–21, 1967.

[JÖR 68] JÖRESKOG K.G., LAWLEY D.N., "New methods in maximum likelihood factor analysis", *British Journal of Mathematical and Statistical Psychology*, vol. 21, pp. 86–96, 1968.

[JÖR 69] JÖRESKOG K.G., "A general approach to confirmatory maximum likelihood factor analysis", *Psychometrika*, vol. 34, pp. 183–202, 1969.

[JÖR 70a] JÖRESKOG K.G., "A general method for analysis of covariance structures", *Biometrika*, vol. 57, pp. 239–251, 1970.

[JÖR 70b] JÖRESKOG K.G., GRUVAEUS G.T., VAN THILLO M., "ACOVS: a general computer program for analysis of covariance structures", *ETS Research Bulletin Series*, pp. 1–54, 1970.

[JÖR 71] JÖRESKOG K.G., VAN THILLO M., "New rapid algorithms for factor analysis by unweighted least squares, generalized least squares and maximum likelihood", *Research Memorandum*, pp. 71–75, 1971.

[JÖR 73] JÖRESKOG K.G., "Analysis of covariance structures", in KRISHNAIAH P.R. (ed.), *Multivariate analysis-III*, Academic Press, New York, 1973.

[JÖR 74] JÖRESKOG K.G., SÖRBOM D., LISREL III [Computer Software], Scientific Software International, Chicago, 1974.

[JÖR 89] JÖRESKOG K.G., SÖRBOM D., LISREL 7: A Guide to the Program and Applications, Scientific Software International, Chicago, 1989.

[JÖR 93] JÖRESKOG K.G., SÖRBOM D., LISREL 8: Structural Equation Modeling with the SIMPLIS Command Language, Scientific Software International, Chicago, 1993.

[KAP 01] KAPLAN D., HARIK P., HOTCHKISS L., "Cross-sectional estimation of dynamic structural equation models in disequilibrium", in CUDECK R., JÖRESKOG K.G., DUTOIT S.H.C. *et al.* (eds), *Structural Equation Modeling: Present and Future*, Scientific Software International, Lincolnwood, 2001.

[KEN 95] KENNY D.A., ZAUTRA A., "The trait-state-error model for multiwave data", *Journal of Consulting and Clinical Psychology*, vol. 63, no. 1, pp. 52–59, 1995.

[KEN 01] KENNY D.A., ZAUTRA A., "Trait-state models for longitudinal data", in COLLINS L.M., SAYER A.G. (eds), *New Methods for the Analysis of Change*, American Psychological Association, Washington, 2001.

[KEN 06] KENNY D.A., KASHY D.A., COOK W.L., *Dyadic Data Analysis*, Guilford Press, New York, 2006.

[KIM 05] KIM K.H., "The relation among fit indexes, power, and sample size in structural equation modeling", *Structural Equation Modeling*, vol. 12, pp. 368–390, 2005.

[KLI 16] KLINE R.B., *Principles and Practice of Structural Equation Modeling*, 4th edition, Guilford Press, New York, 2016.

[KNA 78] KNAPP T.R., "Canonical correlation analysis: a general parametric significance testing system", *Psychological Bulletin*, vol. 85, pp. 410–416, 1978.

[KOR 14] KORKMAZ S., GOKSULUK D., ZARARSIZ G., "MVN: an R package for assessing multivariate normality", *The R Journal*, vol. 6, no. 2, pp. 151–162, 2014.

[LAD 89] LADU T.J., TANAKA J.S., "Influence of sample size, estimation method, and model specification on goodness-of-fit assessments in structural equation models", *Journal of Applied Psychology*, vol. 74, no. 4, pp. 625–635, 1989.

[LAW 40] LAWLEY D.N., "The estimation of factor loadings by the method of maximum likelihood", *Proceedings of the Royal Society of Edinburgh*, vol. 60, pp. 64–82, 1940.

[LAW 43] LAWLEY D.N., "The application of the maximum likelihood method to factor analysis", *British Journal of Psychology*, vol. 33, pp. 172–175, 1943.

[LAZ 55] LAZARSFELD P.F., "Recent developments in latent structure analysis", *Sociometry*, vol. 18, no. 4, pp. 391–403, 1955.

[LEV 02] LEVY B., SLADE M., KUNKEL S., KASL S., "Longevity increased by positive self-perceptions of aging", *Journal of Personality and Social Psychology*, vol. 83, no. 2, pp. 261–270, 2002.

[LI 16] LI C., "The performance of ML, DWLS, and ULS estimation with robust corrections in structural equation models with ordinal variables", *Psychological Methods*, vol. 21, no. 3, pp. 369–387, 2016.

[LIT 02] LITTLE T.D., CUNNINGHAM W.A., SHAHAR G., WIDAMAN K.F., "To parcel or not to parcel: exploring the question, weighing the merits", *Structural Equation Modeling*, vol. 9, pp. 151–173, 2002.

[LIT 13] LITTLE T.D., *Longitudinal Structural Equation Modeling*, Guilford Press, New York, 2013.

[LIU 17] LIU Y., MILLSAP R.E., WEST S.G., TEIN J., TANAKA R., GRIMM K.J., "Testing measurement invariance in longitudinal data with ordered-categorical measures", *Psychological Methods*, vol. 22, no. 3, pp. 486–506, 2017.

[LOE 98] LOEHLIN J.C., *Latent Variable Models. An Introduction to Factor, Path, and Structural Analysis*, 3rd edition, Lawrence Erlbaum Associates Publishers, Mahwah, 1998.

[LOE 04] LOEHLIN J.C., *Latent Variable Models: An Introduction to Factor, Path, and Structural Equation Analysis*, 4th edition, Lawrence Erlbaum Associates Publishers, Mahwah, 2004.

[MAC 86] MACCALLUM R., "Specification searches in covariance structure modeling", *Psychological Bulletin*, vol. 100, no. 1, pp. 107–120, 1986.

[MAC 92] MACCALLUM R.C., ROZNOWSKI M., NECOWITZ L.B., "Model modifications in covariance structure analysis: the problem of capitalization on chance", *Psychological Bulletin*, vol. 111, no. 3, pp. 490–504, 1992.

[MAC 95] MACCALLUM R.C., "Model specification: procedures, strategies, and related issues", in HOYLE R.H. (ed.), *Structural Equation Modeling: Concepts, Issues, and Applications*, Sage Publications, Thousand Oaks, 1995.

[MAC 96] MACCALLUM R.C., BROWNE M.W., SUGAWARA H.M., "Power analysis and determination of sample size for covariance structure modeling", *Psychological Methods*, vol. 1, no. 2, pp. 130–149, 1996.

[MAC 02] MACKINNON D.P., LOCKWOOD C.M., HOFFMAN J.M. *et al.*, "A comparison of methods to test mediation and other intervening variable effects", *Psychological Methods*, vol. 7, no. 1, pp. 83–104, 2002.

[MAC 04] MACKINNON D.P., LOCKWOOD C.M., WILLIAMS J., "Confidence limits for the indirect effect: distribution of the product and resampling methods", *Multivariate Behavioral Research*, vol. 39, no. 1, pp. 99–128, 2004.

[MAC 06] MacCallum R.C., Browne M.W., Cai L., "Testing differences between nested covariance structure models: power analysis and null hypotheses", *Psychological Methods*, vol. 11, no. 1, pp. 19–35, 2006.

[MAC 08] MacKinnon D.P., *Introduction to Statistical Mediation Analysis*, Taylor & Francis Group/Lawrence Erlbaum Associates Publishers, New York, 2008.

[MAL 64] Malinvaud E., *Méthodes statistiques de l'économétrie*, Dunod, Paris, 1964.

[MAR 70] Mardia K.V., "Measures of multivariate skewness and kurtosis with applications", *Biometrika*, vol. 57, pp. 519–530, 1970.

[MAR 85] Marsh H.W., Hocevar D., "Application of confirmatory factor analysis to the study of self-concept: first- and higher order factor models and their invariance across groups", *Psychological Bulletin*, vol. 97, pp. 562–582, 1985.

[MAR 88] Marsh H.W., Balla J.R., McDonald R.P., "Goodness-of-fit indexes in confirmatory factor analysis: the effect of sample size", *Psychological Bulletin*, vol. 103, no. 3, pp. 391–410, 1988.

[MAR 98] Maruyama M.G., *Basics of Structural Equation Modeling*, Sage Publications, London, 1998.

[MAR 13] Marsh H.W., Lüdtke O., Nagengast B. *et al.*, "Why item parcels are (almost) never appropriate: two wrongs do not make a right – Camouflaging misspecification with item parcels in CFA models", *Psychological Methods*, vol. 18, no. 3, pp. 257–284, 2013.

[MAS 91] Mason C.H., Perreault Jr W.D., "Collinearity, power, and interpretation of multiple regression analysis", *Journal of Marketing Research*, vol. 28, pp. 268–280, 1991.

[MAT 12] Matsueda R.L., "Key advances in the history of structural equation modeling", in Hoyle R.H. (ed.), *Handbook of Structural Equation Modeling*, Guilford Press, New York, 2012.

[MCA 07] McArdle J.J., "Dynamic structural equation modeling in longitudinal experimental studies", in van Montfort K., Oud J., Satorra A. (eds), *Longitudinal Models in the Behavioral and Related Sciences*, Lawrence Erlbaum Associates Publishers, Mahwah, 2007.

[MCA 14] McArdle J.J., Nesselroade J.R., *Longitudinal Data Analysis Using Structural Equation Models*, American Psychological Association, Washington, 2014.

[MCD 99] McDonald R.P., *Test Theory: A Unified Treatment*, Lawrence Erlbaum Associates Publishers, Mahwah, 1999.

[MCI 12] McIntosh A.R., "Tracing the route to path analysis in neuroimaging", *Neuroimage*, vol. 62, no. 2, pp. 887–890, 2012.

[MIC 89] Micceri T., "The unicorn, the normal curve, and other improbable creatures", *Psychological Bulletin*, vol. 105, no. 1, pp. 156–166, 1989.

[MOO 93] MOOIJAART A., "Structural equation models with transformed variables", in HAAGEN K., BARTHOLOMEW D.J., DEISTLER M. (eds), *Statistical Modeling and Latent Variables*, Elsevier, Amsterdam, 1993.

[MOR 94] MORROW-HOWELL N., "The M word: multicollinearity in multiple regression", *Social Work Research*, vol. 18, no. 4, pp. 247–251, 1994.

[MUE 96] MUELLER R.O., *Basic Principles of Structural Equation Modeling: An Introduction to LISREL and EQS*, Springer-Verlag, New York, 1996.

[MUL 89] MULAIK S.A., JAMES L.R., VAN ALSTINE J. *et al.*, "Evaluation of goodness-of-fit indices for structural equation models", *Psychological Bulletin*, vol. 105, pp. 430–445, 1989.

[MUL 95] MULAIK S.A., JAMES L.R., "Objectivity and reasoning in science and structural equation modeling", in HOYLE R.H. (ed.), *Structural Equation Modeling: Concepts, Issues, and Applications*, Sage Publications, Thousand Oaks, 1995.

[MUT 83] MUTHÉN B.O., "Latent variable structural equation modeling with categorical data", *Journal of Econometrics*, vol. 22, pp. 43–65, 1983.

[MUT 84] MUTHÉN B.O., "A general structural equation model with dichotomous, ordered categorical, and continuous latent variable indicators", *Psychometrika*, vol. 49, no. 1, pp. 115–132, 1984.

[MUT 87] MUTHÉN B.O., "Response to Freedman's critique of path analysis: improve credibility by better methodological training", *Journal of Educational Statistics*, vol. 12, pp. 178–184, 1987.

[MUT 93] MUTHÉN B.O., "Goodness of fit with categorical and other non-normal variables", in BOLLEN K., LONG S. (eds), *Testing Structural Equation Models*, Sage Publications, Newbury Park, 1993.

[MUT 99] MUTHÉN B.O., SHEDDEN K., "Finite mixture modeling with mixture outcomes using the EM algorithm", *Biometrics*, vol. 55, pp. 463–469, 1999.

[MUT 01] MUTHÉN B.O., Mplus Discussion, accessed 17 April 2017, available at: http://www.statmodel.com/discussion/messages/14/112.html?1376660789, 2001.

[MUT 02a] MUTHÉN B.O., "Beyond SEM: General latent variable modeling", *Behavior Metrika*, vol. 29, no. 1, pp. 81–117, 2002.

[MUT 02b] MUTHÉN L.K., MUTHÉN B.O., "How to use a Monte Carlo study to decide on sample size and determine power", *Structural Equation Modeling*, vol. 9, no. 4, pp. 599–620, 2002.

[MUT 06] MUTHÉN L.K., MUTHÉN B.O., *Mplus User's Guide*, 4th edition, Muthén & Muthén, Los Angeles, 1998–2006.

[NAG 99] NAGIN D.S., "Analyzing developmental trajectories: a semi-parametric, group-based approach", *Psychological Methods*, vol. 4, pp. 139–157, 1999.

[NEW 15] NEWSOM J.T., *Longitudinal Structural Equation Modeling: A Comprehensive Introduction*, Routledge/Taylor & Francis Group, New York, 2015.

[NIL 22] NILES H.E., "Correlation, causation and Wright's theory of path coefficients", *Genetics*, vol. 7, pp. 258–273, 1922.

[NIL 23] NILES H.E., "The method of path coefficients: an answer to Wright", *Genetics*, vol. 7, pp. 256–260, 1923.

[PAU 85] PAULRE B., *La Causalité en économie : signification et portée de la modélisation structurelle*, PUL, Lyon, 1985.

[PEA 36] PEARSON K., "Method of moments and method of maximum likelihood", *Biometrika*, vol. 28, pp. 34–59, 1936.

[PEA 98] PEARL J., "Graphs, causality and structural equation models", *Sociological Methods and Research*, vol. 27, no. 2, pp. 226–284, 1998.

[PET 13] PETRESCU M., "Marketing research using single-item indicators in structural equation models", *Journal of Marketing Analytics*, vol. 1, pp. 99–117, 2013.

[PIE 75] PIERON H., "Préface", in REUCHIN M. (ed.), *Les Méthodes quantitatives en psychologie*, 2nd edition, PUF, Paris, 1975.

[PLO 04] PLOYHART R.E., OSWALD F.L., "Applications of mean and covariance structure analysis: integrating correlational and experimental approaches", *Organizational Research Methods*, vol. 7, pp. 27–65, 2004.

[PRE 06] PREACHER K.J., "Quantifying parsimony in structural equation modeling", *Multivariate Behavioral Research*, vol. 41, no. 3, pp. 227–259, 2006.

[PRE 08a] PREACHER K.J., HAYES A.F., "Asymptotic and resampling strategies for assessing and comparing indirect effects in multiple mediator models", *Behavior Research Methods*, vol. 40, no. 3, pp. 879–891, 2008.

[PRE 08b] PREACHER K.J., WICHMAN A.L., MACCALLUM R.C. *et al.*, *Latent Growth Curve Modeling*, Sage Publications, London, 2008.

[RAM 13] RAMIREZ E., DAVID M.E., BRUSCO M.J., "Marketing's SEM based nomological network: constructs and research streams in 1987–1997 and in 1998–2008", *Journal of Business Research*, vol. 66, no. 9, pp. 1255–1260, 2013.

[RAY 01a] RAYKOV T., "Estimation of congeneric scale reliability using covariance structure analysis with nonlinear constraints", *British Journal of Mathematical and Statistical Psychology*, vol. 54, pp. 315–323, 2001.

[RAY 01b] RAYKOV T., PENEV S., "The problem of equivalent structural equation models: an individual residual perspective", in MARCOULIDES G.A., SCHUMACKER R.E. (eds), *New Developments and Techniques in Structural Equation Modeling*, Lawrence Erlbaum Associates Publishers, Mahwah, 2001.

[REN 98] RENSVOLD R.B., CHEUNG G.W., "Testing measurement model for factorial invariance: a systematic approach", *Educational and Psychological Measurement*, vol. 58, no. 6, pp. 1017–1034, 1998.

[REU 62] REUCHLIN M., *Les Méthodes quantitatives en psychologie*, PUF, Paris, 1962.

[REU 64] REUCHLIN M., *Méthodes d'analyse factorielle à l'usage des psychologues*, PUF, Paris, 1964.

[REU 95] REUCHLIN M., *Totalités, éléments, structures en psychologie*, PUF, Paris, 1995.

[RIG 95] RIGDON E.E., "A necessary and sufficient identification rule for structural models estimated in practice", *Multivariate Behavioral Research*, vol. 30, no. 3, pp. 359–383, 1995.

[ROS 65] ROSENBERG M., *Society and the Adolescent Self-Image*, Princeton University Press, Princeton, 1965.

[ROS 12] ROSSEEL Y., "Lavaan: an R package for structural equation modeling", *Journal of Statistical Software*, vol. 48, no. 2, pp. 1–36, 2012.

[ROU 02] ROUSSEL P., COMPOY E., DURRIEU F., *Méthodes d'équations structurelles : recherche et applications en gestion*, Economica, Paris, 2002.

[RUB 95] RUBIO D.M., GILLESPIE G.F., "Problems with error in structural equation models", *Structural equation modeling*, vol. 2, pp. 367–378, 1995.

[RUS 47] RUSSELL B., *L'Esprit scientifique et la science dans le monde moderne*, Janin, Paris, 1947.

[SAS 11] SAS INSTITUTE INC., *SAS/STAT® 9.3 User's Guide*, SAS Institute Inc, Cary, 2011.

[SAT 01] SATORRA A., BENTLER P.M., "A scaled difference chi-square test statistic for moment structure analysis", *Psychometrika*, vol. 66, no. 4, pp. 507–514, 2001.

[SCH 94] SCHEIER M.F., CARVER C.S., BRIDGES M.W., "Distinguishing optimism from neuroticism (and trait anxiety, self-mastery, and self-esteem): a reevaluation of the life orientation test", *Journal of Personality and Social Psychology*, vol. 67, no. 6, pp. 1063–1078, 1994.

[SCH 06] SCHREIBER J.B., STAGE F.K., KING J. *et al.*, "Reporting structural equation modeling and confirmatory factor analysis results: a review", *The Journal of Educational Research*, vol. 99, no. 6, pp. 323–337, 2006.

[SIM 53] SIMON H.A., "Causal ordering and identifiability", in HOOD W.C., KOOPMANS T.C. (eds), *Studies in Econometric Method. Cowles Commission for Research in Economics, Monograph No. 14*, John Wiley & Sons, New York, 1953.

[SIM 54] SIMON H.A., "Spurious correlation: a causal interpretation", *Journal of the American Statistical Association*, vol. 49, pp. 467–479, 1954.

[SIN 03] SINGER J.D., WILLETT J.B., *Applied Longitudinal Data Analysis: Modeling Change and Event Occurrence*, Oxford University Press, New York, 2003.

[SNY 91] SNYDER C.R., HARRIS C., ANDERSON J.R. *et al.*, "The will and the ways: development and validation of an individual differences measure of hope", *Journal of Personality and Social Psychology*, vol. 60, pp. 570–585, 1991.

[SOB 82] SOBEL M.E., "Asymptotic intervals for indirect effects in structural equations models", in LEINHART S. (ed.), *Sociological Methodology*, Jossey-Bass, San Francisco, 1982.

[SOB 87] SOBEL M.E., "Direct and indirect effects in linear structural equation models", *Sociological Methods and Research*, vol. 16, pp. 155–176, 1987.

[SÖR 01] SÖRBOM D., "Karl Jöreskog and LISREL: a personal story", in CUDECK R., DUTOIT S., SÖRBOM D. (eds), *Structural Equation Modeling: Present and Future – A Festschrift in Honor of Karl G Jöreskog*, Scientific Software International, Lincolnwood, 2001.

[SPE 04] SPEARMAN C., "General intelligence: objectively determined and measured", *American Journal of Psychology*, vol. 5, pp. 201–293, 1904.

[STE 15] STEYER R., MAYER A., GEISER C. *et al.*, "A theory of states and traits – Revised", *Annual Review of Clinical Psychology*, vol. 11, pp. 71–98, 2015.

[STI 07] STIGLER S.M., "The epic story of maximum likelihood", *Statistical Science*, vol. 22, pp. 598–620, 2007.

[STO 04] STOEL R.D., VANDEN WITTENBOER G., HOX J.J., "Methodological issues in the application of the latent growth curve model", in VAN MONTFORT K., OUD H., SATORRA A. (eds), *Recent Developments on Structural Equation Modeling: Theory and Applications*, Kluwer Academic Press, Amsterdam, 2004.

[SUY 88] SUYAPA E., SILVA M., MACCALLUM R.C., "Some factors affecting the success of specification searches in covariance structure modeling", *Multivariate Behavioral Research*, vol. 23, no. 3, pp. 297–326, 1988.

[TAB 07] TABACHNICK B.G., FIDELL L.S., *Using Multivariate Statistics*, 5th edition, Allyn & Bacon/Pearson Education, Boston, 2007.

[TAN 93] TANAKA J.S., "Multifaceted conceptions of fit in structure equation models", in BOLLEN K.A., LONG J.S. (eds), *Testing Structural Equation Models*, Sage, Newbury Park, 1993.

[THO 13] THORNDIKE E.L., *Educational Psychology, Vol 2: The Psychology of Learning*, Teachers College, New York, 1913.

[TOL 38] TOLMAN E.C., "The determiners of behavior at a choice point", *Psychological Review*, vol. 45, no. 1, pp. 1–41, 1938.

[TUC 58] TUCKER L.R., "Determination of parameters of a functional relation by factor analysis", *Psychometrika*, vol. 23, pp. 19–23, 1958.

[TUC 66] TUCKER L.R., "Some mathematical notes on three-mode factor analysis", *Psychometrika*, vol. 31, pp. 279–311, 1966.

[WER 70] WERTS C.E., LINN R.L., "Path analysis: psychological examples", *Psychological Bulletin*, vol. 74, pp. 193–212, 1970.

[WES 95] WEST S.G., FINCH J.F., CURRAN P.J., "Structural equation models with nonnormal variables: problems and remedies", in HOYLE R.H. (ed.), *Structural Equation Modeling: Concepts, Issues, and Applications*, Sage Publications, Thousand Oaks, 1995.

[WES 10] WESTLAND J.C., "Lower bounds on sample size in structural equation modeling", *Electronic Commerce Research and Applications*, vol. 9, no. 6, pp. 476–487, 2010.

[WIL 99] WILKINSON L., "Statistical methods in psychology journals: guidelines and explanations", *American Psychologist*, vol. 54, no. 8, pp. 594–604, 1999.

[WIL 11] WILLIAMS L.J., O'BOYLE E.J., "The myth of global fit indices and alternatives for assessing latent variable relations", *Organizational Research Methods*, vol. 14, no. 2, pp. 350–369, 2011.

[WIL 12] WILLIAMS L.J., "Equivalent models: Concepts, problems, alternatives", in HOYLE R.H. (ed.), *Handbook of Structural Equation Modeling*, Guilford Press, New York, 2012.

[WOL 13] WOLF E.J., HARRINGTON K.M., CLARK S.L., MILLER M.W., "Sample size requirements for structural equation models: an evaluation of power, bias, and solution propriety", *Educational and Psychological Measurement*, vol. 76, no. 6, pp. 913–934, 2013.

[WON 99] WONG C., LAW K.S., "Testing reciprocal relations by non-recursive structural equation models using cross-sectional data", *Organizational Research Methods*, vol. 2, no. 1, pp. 69–87, 1999.

[WOT 93] WOTHKE W., "Nonpositive definite matrices in structural modeling", in BOLLEN K.A., LONG J.S. (eds), *Testing Structural Equation Models*, Sage, Newbury Park, 1993.

[WRI 21a] WRIGHT S., "Correlation and causation", *Journal of Agricultural Research*, vol. 20, pp. 557–585, 1921.

[WRI 21b] WRIGHT S., "Systems of mating I-V", *Genetics*, vol. 6, pp. 111–178, 1921.

[WRI 34] WRIGHT S., "The method of path coefficients", *Annals of Mathematical Statistics*, vol. 5, pp. 161–215, 1934.

[WRI 60] WRIGHT S., "Path coefficients and path regressions: alternative or complementary concepts?", *Biometrics*, vol. 16, pp. 189–202, 1960.

[YUA 00] YUAN K., CHAN W., BENTLER P.M., "Robust transformation with applications to structural equation modelling", *British Journal of Mathematical and Statistical Psychology*, vol. 53, no. 1, pp. 31–50, 2000.

[YUN 99] YUNG Y.F., THISSEN D., MCLEOD L.D., "On the relationship between the higher-order factor model and the hierarchical factor model", *Psychometrika*, vol. 64, no. 2, pp. 113–128, 1999.

[ZHA 15] ZHANG Z., HAMAGAMI F., GRIMM K.J. *et al.*, "Using R package RAMpath for tracing SEM path diagrams and conducting complex longitudinal data analysis", *Structural Equation Modeling*, vol. 22, no. 1, pp. 132–147, 2015.

[ZHA 17] ZHANG Z., "Structural equation modeling in the context of clinical research", *Annals of Translational Medicine*, vol. 5, no. 5, pp. 102, 2017.

Index

A, B

C, D

E, F, H

I, J

K, L, M

N, O, P

Q, R, S

T, U

V, W

Other titles from

ISTE

in

Mathematics and Statistics

2018

AZAÏS Romain, BOUGUET Florian
Statistical Inference for Piecewise-deterministic Markov Processes

IBRAHIMI Mohammed
Mergers & Acquisitions: Theory, Strategy, Finance

PARROCHIA Daniel
Mathematics and Philosophy

2017

CARONI Chysseis
First Hitting Time Regression Models: Lifetime Data Analysis Based on Underlying Stochastic Processes
(Mathematical Models and Methods in Reliability Set – Volume 4)

CELANT Giorgio, BRONIATOWSKI Michel
Interpolation and Extrapolation Optimal Designs 2: Finite Dimensional General Models

CONSOLE Rodolfo, MURRU Maura, FALCONE Giuseppe
Earthquake Occurrence: Short- and Long-term Models and their Validation
(Statistical Methods for Earthquakes Set – Volume 1)

D'AMICO Guglielmo, DI BIASE Giuseppe, JANSSEN Jacques, MANCA Raimondo
Semi-Markov Migration Models for Credit Risk
(Stochastic Models for Insurance Set – Volume 1)

GONZÁLEZ VELASCO Miguel, del PUERTO GARCÍA Inés, YANEV George P.
Controlled Branching Processes
(Branching Processes, Branching Random Walks and Branching Particle Fields Set – Volume 2)

HARLAMOV Boris
Stochastic Analysis of Risk and Management
(Stochastic Models in Survival Analysis and Reliability Set – Volume 2)

KERSTING Götz, VATUTIN Vladimir
Discrete Time Branching Processes in Random Environment
(Branching Processes, Branching Random Walks and Branching Particle Fields Set – Volume 1)

MISHURA YULIYA, SHEVCHENKO Georgiy
Theory and Statistical Applications of Stochastic Processes

NIKULIN Mikhail, CHIMITOVA Ekaterina
Chi-squared Goodness-of-fit Tests for Censored Data
(Stochastic Models in Survival Analysis and Reliability Set – Volume 3)

SIMON Jacques
Banach, Fréchet, Hilbert and Neumann Spaces
(Analysis for PDEs Set – Volume 1)

2016

CELANT Giorgio, BRONIATOWSKI Michel
Interpolation and Extrapolation Optimal Designs 1: Polynomial Regression and Approximation Theory

CHIASSERINI Carla Fabiana, GRIBAUDO Marco, MANINI Daniele
Analytical Modeling of Wireless Communication Systems
(Stochastic Models in Computer Science and Telecommunication Networks Set – Volume 1)

GOUDON Thierry
Mathematics for Modeling and Scientific Computing

KAHLE Waltraud, MERCIER Sophie, PAROISSIN Christian
Degradation Processes in Reliability
(Mathematial Models and Methods in Reliability Set – Volume 3)

KERN Michel
Numerical Methods for Inverse Problems

RYKOV Vladimir
Reliability of Engineering Systems and Technological Risks
(Stochastic Models in Survival Analysis and Reliability Set – Volume 1)

2015

DE SAPORTA Benoîte, DUFOUR François, ZHANG Huilong
Numerical Methods for Simulation and Optimization of Piecewise Deterministic Markov Processes

DEVOLDER Pierre, JANSSEN Jacques, MANCA Raimondo
Basic Stochastic Processes

LE GAT Yves
Recurrent Event Modeling Based on the Yule Process
(Mathematical Models and Methods in Reliability Set – Volume 2)

2014

COOKE Roger M., NIEBOER Daan, MISIEWICZ Jolanta
Fat-tailed Distributions: Data, Diagnostics and Dependence
(Mathematical Models and Methods in Reliability Set – Volume 1)

MACKEVIČIUS Vigirdas
Integral and Measure: From Rather Simple to Rather Complex

PASCHOS Vangelis Th
Combinatorial Optimization – 3-volume series – 2nd edition
Concepts of Combinatorial Optimization / Concepts and Fundamentals – volume 1
Paradigms of Combinatorial Optimization – volume 2
Applications of Combinatorial Optimization – volume 3

2013

COUALLIER Vincent, GERVILLE-RÉACHE Léo, HUBER Catherine, LIMNIOS Nikolaos, MESBAH Mounir
Statistical Models and Methods for Reliability and Survival Analysis

JANSSEN Jacques, MANCA Oronzio, MANCA Raimondo
Applied Diffusion Processes from Engineering to Finance

SERICOLA Bruno
Markov Chains: Theory, Algorithms and Applications

2012

BOSQ Denis
Mathematical Statistics and Stochastic Processes

CHRISTENSEN Karl Bang, KREINER Svend, MESBAH Mounir
Rasch Models in Health

DEVOLDER Pierre, JANSSEN Jacques, MANCA Raimondo
Stochastic Methods for Pension Funds

2011

MACKEVIČIUS Vigirdas
Introduction to Stochastic Analysis: Integrals and Differential Equations

MAHJOUB Ridha

Recent Progress in Combinatorial Optimization – ISCO2010

RAYNAUD Hervé, ARROW Kenneth
Managerial Logic

2010

BAGDONAVIČIUS Vilijandas, KRUOPIS Julius, NIKULIN Mikhail
Nonparametric Tests for Censored Data

BAGDONAVIČIUS Vilijandas, KRUOPIS Julius, NIKULIN Mikhail
Nonparametric Tests for Complete Data

IOSIFESCU Marius *et al.*
Introduction to Stochastic Models

VASSILIOU PCG
Discrete-time Asset Pricing Models in Applied Stochastic Finance

2008

ANISIMOV Vladimir
Switching Processes in Queuing Models

FICHE Georges, HÉBUTERNE Gérard
Mathematics for Engineers

HUBER Catherine, LIMNIOS Nikolaos *et al.*
Mathematical Methods in Survival Analysis, Reliability and Quality of Life

JANSSEN Jacques, MANCA Raimondo, VOLPE Ernesto
Mathematical Finance

2007

HARLAMOV Boris
Continuous Semi-Markov Processes

2006

CLERC Maurice
Particle Swarm Optimization

Printed and bound by CPI Group (UK) Ltd, Croydon, CR0 4YY

15/09/2026

14962351-0005